C000174941

LANDSCAPE OF THE MIND

HUMAN EVOLUTION AND THE

ARCHAEOLOGY OF THOUGHT

LANDSCAPE
OF THE MIND

JOHN F. HOFFECKER

COLUMBIA UNIVERSITY PRESS NEW YORK

Columbia University Press
Publishers Since 1893
New York   Chichester, West Sussex

Copyright © 2011 John F. Hoffecker
All rights reserved

Library of Congress Cataloging-in-Publication Data
Hoffecker, John F.
Landscape of the mind : human evolution and the archaeology of thought / John F. Hoffecker.
p. cm.
Includes bibliographical references and index.
ISBN 978-0-231-14704-0 (cloth : alk. paper)
ISBN 978-0-231-51848-2 (e-book)
1. Human evolution. 2. Brain—Evolution. 3. Thought and thinking.
I. Title.

GN281.H62   2011
153.4—dc22   2010045026

Columbia University Press books are printed on permanent and durable acid-free paper.
This book is printed on paper with recycled content.
Printed in the United States of America
c 10 9 8 7 6 5 4 3 2 1

References to Internet Web sites (URLs) were accurate at the time of writing. Neither the author
nor Columbia University Press is responsible for URLs that may have expired or changed since
the manuscript was prepared.

TO THE MAN AND HIS WORK
V. GORDON CHILDE
(1892–1957)

CONTENTS

PREFACE AND ACKNOWLEDGMENTS

ALTHOUGH READERS NEED NOT concern themselves with the particulars, death was very much on my mind as I wrote *Landscape of the Mind*. More to the point, the bitter gulf that lies between the death of a person as a biological organism and the potential immortality of his or her thoughts was on my mind. It reflects a conviction that "the mind" is an entity that transcends both biological space (that is, an individual brain) and biological time. Another conviction is that we—I refer to not only archaeologists, but also students of other disciplines—have been studying something important that remains only vaguely defined.

If the ultimate source of the mind is the individual brain—an unmistakable product of evolutionary biology—something has emerged with the mind, something that involves properties not found elsewhere in the biological realm. Some refer to them as emergent properties and attribute them to the almost unimaginable complexity of the system they represent. It is a system much greater than the estimated 1 million billion synaptic connections in an individual brain (itself described as the most complex object in the known universe). At some point in human evolution, probably not very long ago, humans evolved the capacity for transmitting thoughts from one brain to others by means of externalized symbols. This created a "super-brain" within each social group and probably marked the advent of both the mind and what many of my colleagues in archaeology refer to as modernity.

The characteristic of the mind most important to our lives is creativity—the capacity for the seemingly limitless recombination of thoughts expressed as words or artifacts into hierarchically organized structures with no prior existence. The first bow and arrow is one example, and *Catcher in the Rye* is another. The advent of creativity changed everything. It brought forth onto the Earth something new, analogous to the origin of life itself. The creative mind has been developing and accumulating these structures ever since. One of the consequences is that most of us spend most of our time in a landscape shaped largely by structures of the mind, not by the processes of geomorphology or evolutionary biology.

Another conviction that readers will find expressed in this book is that archaeology can contribute to an understanding of the mind, including its origin. Although much of archaeological method may seem rather simple and antiquated, the data of the archaeological record is largely a trail of fossil thought. This is a consequence of the ability, very rare among living organisms (the honeybee being an example), to project mental representations outside the brain, which was something accomplished initially with the hands and later with the vocal tract. A central theme of this book is that by reaching out to touch and eventually alter their own thoughts, evolving humans made the mind.

As I wrote this book, I was acutely conscious of how the creativity of others—many of them no longer living—had provided much of the structure of thought contained here. The thoughts of V. Gordon Childe (1892-1957) about archaeology and history are fundamental, as are the ideas of R. G. Collingwood (1889-1943), who was an influence on Childe. Some major ideas that readers will find in this book about the relationship between humans and their artifacts were articulated by André Leroi-Gourhan (1911-1986) several decades ago, although most English-speaking archaeologists are unaware of his contribution in this realm. Noam Chomsky's ideas about language—especially those of the past two decades—are an essential foundation. Much of what I have written about the evolution and spread of modern humans is based on the work of Richard G. Klein.

My parents, John Savin and Felicity Hoffecker, nurtured my interest in archaeology and history from an early age. My father was especially interested in Collingwood's work and often spoke about it, along with that of Benedetto Croce (1856-1952), who was, in turn, an influence on Collingwood. Many of the illustrations were done by Ian Torao Hoffecker. They are all part of this book.

A number of people read and commented on specific chapters of the book, and I am grateful for their time and thoughts. Colleagues and friends who read the critically important (and heavily revised) first chapter include Jolanta M. Grajski, Nancy R. Lyons, Marina Petrova, and Shelly Sommer. The first chapter also reflects fruitful discussions with Bob Levin and Valerie E. Stone.

My colleagues in archaeology at the University of Colorado in Boulder reviewed chapter 4 in the congenial setting of our monthly luncheon, and I would like to thank Cathy Cameron, Art Joyce, Steve Lekson, Paola Villa, and Richard Wilshusen for their thoughts and comments. Three anonymous reviewers for Columbia University Press plowed through the complete draft, and their comments and criticisms were very helpful.

I am especially grateful to the staff at Columbia University Press. Patrick Fitzgerald, Publisher for the Life Sciences, embraced this project with enthusiasm at the outset and remained an important source of guidance and encouragement throughout the two and half years that I labored on the draft. The entire manuscript was edited by Irene Pavitt. Editorial assistant Bridget Flannery-McCoy and her predecessor, Marina Petrova, also were very helpful.

LANDSCAPE OF THE MIND

1

MODERNITY AND INFINITY

[T]he mind is in its own nature immortal.

RENÉ DESCARTES

TOWARD THE END OF HIS MEMOIR, Vladimir Nabokov expressed his frustration at "having developed an infinity of sensation and thought within a finite existence."[1] It was a theme he had turned to in the last lines of *Lolita*: "I am thinking of aurochs and angels, the secret of durable pigments, prophetic sonnets, the refuge of art. And this is the only immortality you and I may share."[2] Nabokov died in 1977 and—like all other forms of organic life—relinquished his conscious sensation of being. But many of his thoughts continue to exist, as they do on this page, and in this way Nabokov transcended his finite existence.

It is a power that almost all humans possess—to transcend their existence as organic beings by communicating thoughts that will endure after they die. The ideas of an individual may be communicated in writing or print or electronic media and, before the invention of writing, by oral tradition. In this way, the seemingly ephemeral thoughts that flow through the brain may endure for centuries or more.

Nonverbal thoughts may be communicated through art and technology. The aurochs mentioned by Nabokov refer to cave paintings of the Upper Paleolithic that were created more than 30,000 years ago.[3] The visual imaginings of the individuals who crafted these paintings are as fresh today as the words written by Nabokov several decades ago. The earliest known examples of thoughts expressed in material form are the bifacial stone tools of the Lower Paleolithic, some of which are 1.7 million years old.[4] Much of the archaeological record, like the historical written record, is a record of thought.

External Thought

[O]ur world is the product of our thoughts.
JAMES DEETZ

The immortality of artifact and art is based on a unique human ability to express complex thoughts outside the brain. Although language comes to mind first as the means by which people articulate their ideas, humans externalize thoughts in a wide variety of media. In addition to spoken and written language, these media include music, painting, dance, gesture, architecture, sculpture, and others. One of the most consequential forms of external thought is *technology*. The earliest humans used and modified natural objects, as do some other animals. But during the past million years, humans have externalized thoughts in the form of technologies that have become increasingly complex and powerful with far-reaching effects on themselves and their environment.

Other animals can communicate their emotional states and simple bits of information, but only humans can project complex structures of thought, or *mental representations*, outside the brain.[5] Like those of other animals, the human brain receives what cognitive psychologists term *natural representations*. A natural representation might be the perceived or remembered visual image of a waterfall or the sound of the waterfall. Humans have not yet developed the technology to externalize a natural representation, but they can create and project artificial (or "semantic") representations, such as a painting or verbal description of the perceived or remembered image of the waterfall.[6] From 1.7 million years ago onward, the archaeological record is filled with artificial representations.

Humans have evolved two specialized organs that are used to communicate or externalize mental representations: the hand and the vocal tract (figure 1.1). The hand apparently evolved first as a means to project thoughts outside the brain, underscoring the seminal role of bipedalism in human origins. Although the primates have had a long history of manipulating objects with their forelimbs, it was upright walking among the earliest humans that freed up the hands to become the highly specialized instruments they are today.[7] By roughly 3.5 million years ago, the australopithecine hand exhibits significant divergence from the ape hand—especially with respect to the length of the thumb—and further changes coincide with an increase in brain volume and the earliest known stone tools, at about 2.5 million years ago. The fully modern hand apparently was present at 1.7 to 1.6 million years ago (that is, at the time that the first bifacial tools appear).[8]

As did the hand, the human vocal tract diverged from the ape pattern. Among modern human adults, the larynx is positioned lower in the neck than it is in apes, and the epiglottis is not in contact with the soft palate. Like the hand, the human vocal tract is a highly specialized instrument for externalizing thought. Reconstructing the evolution of the vocal tract is more difficult, however, than tracing the development of the hand. The vocal tract is composed entirely of soft parts that do not preserve in the fossil record, so it has been reconstructed on the basis of associated bones

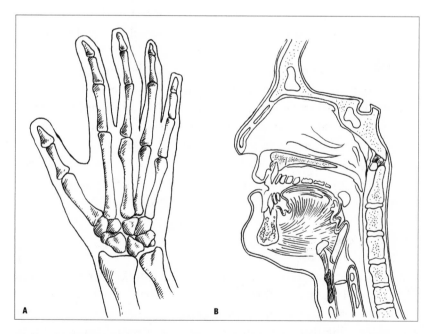

Figure 1.1 Humans evolved two specialized organs to project artificial or semantic mental representations outside the brain. Under fine motor control, both the hand (A) and the vocal tract (B) can execute a potentially infinite array of subtle and sequentially structured movements to transform information in the brain into material objects or symbolically coded sounds. (Drawn by Ian T. Hoffecker)

such as the hyoid. These reconstructions have been controversial, and it remains unclear when the human vocal tract evolved.[9] Furthermore, the relationship between the larynx and language is problematic. In theory, a syntactic language may be produced with a much smaller range of sounds than that offered by the human vocal tract.[10] The critical question is: When did humans begin to create "artifacts" of sound with the vocal tract?

It is the archaeological record, not the human fossil record, that contains the essential clues to the development of external thought. Between about 2.5 and 1.7 million years ago, early forms of the genus *Homo* (and perhaps the last of australopithecines) were making stone tools that, while comparatively simple, were beyond the capacity of the living apes.[11] Not until about 1.7 million years ago did the first recognizable examples of external thought appear, in the form of bifacial tools that exhibit a mental template imposed on rock (figure 1.2).[12] The finished tools bear little resemblance to the original stone fragments from which they were chipped. And it is not until after 500,000 years ago (among both *Homo sapiens* and *Homo neanderthalensis*) that more complex representations began to appear, in the form of composite tools and weapons—assembled from three or four components—reflecting a hierarchical structure.[13]

After 100,000 years ago, evidence for external thought on a scale commensurate with that of living humans finally emerges in the archaeological record: elaborate visual art in both two and three dimensions, remarkably sophisticated musical instruments, personal ornaments, innovative and increasingly complex technology, and other examples.[14] Although spoken language leaves no archaeological traces, its properties are manifest in the art and other media that are preserved. These patterns in the archaeological record, which imply the presence of spoken language, are widely interpreted as the advent of modern behavior, often labeled *modernity*.[15]

Modernity has been defined as "the same cognitive and communication faculties" as those of living humans.[16] This definition has been translated into archaeological terms in at least two ways. The first is a lengthy list of traits that are characteristic of living and recent humans but are not found in the earlier archaeological record. In addition to visual art and ornaments, they include the invasion of previously unoccupied habitats, the production of bone tools, the development of effective hunting techniques, and the organization of domestic space.[17] This approach has been criticized as a "shopping list" that suffers from the

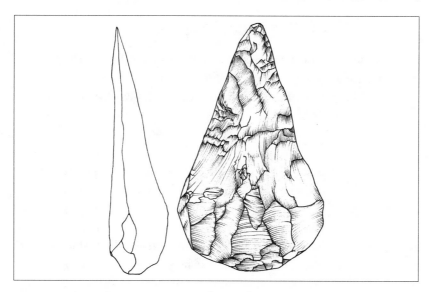

Figure 1.2 The earliest known externalized, or phenotypic, thoughts are objects of stone chipped into a three-dimensional ovate shape in accordance with a mental template. Appearing more than 1.5 million years ago, large bifaces (often termed hand axes) persisted with little change for 1 million years. They seem to reflect an earlier form of mind (or proto-mind) that lacked the unlimited creative potential of the modern mind. (Redrawn by Ian T. Hoffecker, from John Wymer, *Lower Palaeolithic Archaeology in Britain as Represented by the Thames Valley* [London: Baker, 1968], 51, fig. 14)

absence of a theoretical framework regarding the human mind. An alternative archaeological definition of modernity is the "storage of symbols outside the brain" in the form of art, ornamentation, style, and formal spatial patterning.[18]

Both definitions, however, lack a focus on the core property of language and modern behavior in general—the creation of a potentially infinite variety of combinations from a finite set of elements. Living humans externalize thought in a wider range of media than did their predecessors more than 100,000 years ago, but it is the ability to generate an unlimited array of sculptures, paintings, musical compositions, ornaments, tools, and the other media that really sets modern behavior apart from that of earlier humans and all other animals. And it is *creativity*—so clearly manifest in the archaeological record after 100,000 years ago—that provides the firmest basis for inferring the arrival of language.

The Core Property

> To conjure up an internal representation of the future, the brain must have an ability to take certain elements of prior experiences and reconfigure them in a way that in its totality does not correspond to any actual past experience.
>
> ELKHONON GOLDBERG

Noam Chomsky described the core property of language as *discrete infinity* (figure 1.3)[19] Galileo apparently was the first to recognize and appreciate the significance of "the capacity to generate an infinite range of expressions from a finite set of elements"[20] In all languages, people use "rules" of syntax[21] to construct sentences and longer narratives—all of which are artificial representations—out of words that have, in turn, been constructed from a modest assortment of sounds. There seems to be no limit to the variety of sentences and narratives that may be created by individuals who speak a syntactic language.

An essential feature of discrete infinity is hierarchical structure. A syntactic language allows the speaker to create sentences with hierarchically

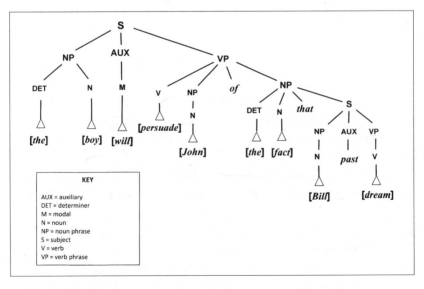

Figure 1.3 The hierarchical structure of a sentence in the English language, illustrated by a tree representation, permits the generation of a potentially unlimited array of expressions. (Redrawn from Noam Chomsky, *Language and Mind*, 3rd ed. [Cambridge: Cambridge University Press, 2006], 129)

organized phrases and subordinate clauses. Most words are themselves composed of smaller units. Without this hierarchical structure, language would be severely constrained. Experiments with nonhuman primates (for example, tamarins) have shown that while some animals can combine and recombine a small number of symbols only humans can generate an immense lexicon of words and a potentially infinite variety of complex sentences through *phrase structure grammar*.[22] The same principle applies to the genetic code, which yields a potentially infinite variety of structures (phenotypes) through hierarchically organized combinations of an even smaller number of elements (DNA base pairs).[23]

As the evolutionary psychologist Michael Corballis noted several years ago, the core property of language may be applied to all the various media through which humans externalize thoughts, or mental representations. He described the ability to generate a potentially infinite variety of hierarchically organized structures as "the key to the human mind." This principle of creative recombination of elements is manifest, for example, in music, dance, sculpture, cooking recipes, personal ornamentation, architecture, and painting.[24] Among most media, creativity may be expressed not only in digital form (that is, using discrete elements such as sounds and words), but also in analogical or continuous form, such as the curve of a Roman arch or the movements of a ballerina.

The medium of creative expression with the most profound consequences for humankind is technology. The ability to generate novel kinds of technology—externalized mental representations in the form of instruments or facilities used to manipulate the environment as an extension of the body—has had an enormous impact on individual lives and population histories. Most of human technology exhibits a design—some of it highly complex—based on nongenetic information. Humans have redesigned themselves as organic beings with everything from clothing stitched together from cut pieces of animal hide to eyeglasses to electronic pacemakers. They have also redesigned the environment, both abiotic and biotic, in ways that have completely altered their relationship to it as organic beings.

During the Upper Paleolithic (roughly 50,000 to 12,000 years ago), designed technology unfolded on a relatively modest scale with throwing darts, small-mammal traps, fishing nets, baking ovens, and other simple devices for acquiring, processing, and storing food. But widespread domestication (that is, redesigning) of plants and animals in the millennia that followed dramatically altered the relationship between humans and their modified environment. Population density increased during the later phases of the Upper Paleolithic and exploded in the postglacial epoch as villages and cities (and new forms of society) emerged. All these developments were tied to the generation of novel technologies, including those that became increasingly complex. While a spoken sentence composed of a million words is theoretically possible, the constraints of short-term memory and social relationships render it impractical and undesirable. Such biological constraints do not always apply to technology, which may be hierarchically organized on an immense scale like the Internet (composed of billions of elements) (figure 1.4).

By the mid-twentieth century, creativity had become a major topic in psychology, and numerous books and papers were written about it. Psychologists define it in much the same way as linguists define the core property of language: "The forming of associative elements into new combinations which either meet specified requirements or are in some way useful."[25] Beyond the everyday creative use of language, they distinguish various forms of creativity. "Exploratory creativity" entails coming up with a new idea within an established domain of thought, such as sculpture, chess, or clothing design. Another form is "making unfamiliar combinations of familiar ideas," such as the above-noted comparison between the genetic code and language. A rarer form is "transformational," which Margaret Boden described as people "thinking something which, with respect to the conceptual spaces in their minds, they *couldn't* have thought before."[26] An example would be Darwin's theory of natural selection.

As with external thought, the archaeological record reveals when and where creativity emerged in the course of human evolution. Unlike externalized mental representations, however, creativity cannot be recognized in a single artifact or feature in that record. It is only by comparison of artifacts and features in space and time that creativity can be observed in archaeological remains. The earliest externalized representations—large bifacial stone tools of the Lower Paleolithic—exhibit an almost total lack of design variation across time and space. But after 500,000 years ago, a limited degree of creativity is evident in the variations of small bifacial tools, core reduction sequences (that is, the succession of steps by which a piece of stone was broken down to smaller pieces to be used as blanks for tools), and, perhaps especially, the composite tools and weapons that appear by 250,000 years ago.[27]

The unlimited creative potential of living humans is not visible in the archaeological record until after the emergence of anatomically modern humans in Africa. And although remains of anatomically modern humans have been dated in Africa to roughly 200,000 years ago, it is not until some-

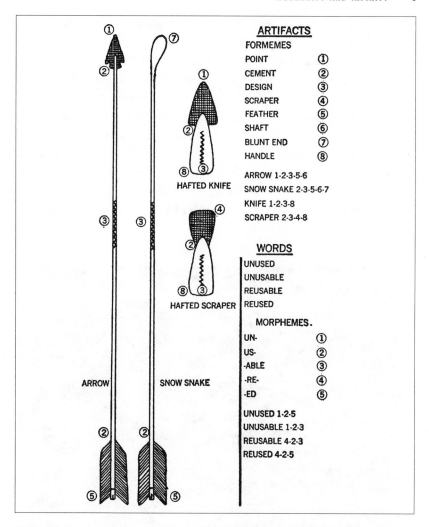

Figure 1.4 Humans can create a potentially infinite variety of tools and other technology, structured hierarchically and much of it in analogical, rather than digital, form. Technology is the means by which the mind engages the external world. (Illustration by Eric G. Engstrom, from James Deetz, *Invitation to Archaeology* [Garden City, N.Y.: Natural History Press, 1967], 91, fig. 16. Reprinted with permission from Patricia Scott Deetz)

what later—after 100,000 years ago and perhaps not until after 50,000 years ago—that creativity on a scale commensurate with that of living humans is evident in the archaeological record. The most vivid illustrations are to be found among the examples of Upper Paleolithic visual art that date to between 35,000 and 30,000 years ago. They include two-dimensional polychrome paintings and small three-dimensional sculptures that exhibit a conceivably unlimited recombination of elements within a complex hierarchical frame-work with embedded components (figure 1.5). Although expressed primar-ily in analogical rather than digital form, each work of art is like a narrative of words and sentences with the potential for infinite variation.[28]

The creativity of the modern human mind is manifest in all the media of external thought that are preserved in the archaeological record of this period. The same pattern illustrated by visual art may be inferred from the presence of musical instruments of the early Upper Paleolithic, but in this case the compositions are lost—only the paint brushes have sur-vived.[29] In the case of technology, however, the creative recombination of elements—and the increasingly complex hierarchical structure of tools

Figure 1.5 The earliest known examples of visual art date to the early Upper Paleolithic, more than 40,000 years ago. Each work of art is unique and is hierarchically structured like the representations generated with syntactic language. This female figurine was recently recovered from Hohle Fels Cave (southwestern Germany). (From Nicholas J. Conard, "A Female Figurine from the Basal Aurignacian of Hohle Fels Cave in South-western Germany," Nature 459 [2009]: 248, fig. 1. Reprinted by permission from Macmil-lan Publishers Ltd. Nature, copyright 2009)

and other means for effecting changes on the world—is readily apparent from 50,000 years ago onward. Novel technologies include sewn clothing, heated artificial shelters, watercraft, and devices for harvesting small game and fish and/or waterfowl. By 25,000 years ago, there is evidence for winter houses, fired ceramics, refrigerated storage, and portable lamps, and during the late Upper Paleolithic (about 20,000 to 12,000 years ago), for domesticated animals (dogs) and mechanical devices.[30]

It is creativity in the archaeological record that signals the advent of modernity, not the appearance of symbols (or the "storage of symbols" in artifacts and other material forms). Symbols in the strict sense are not especially common in the Upper Paleolithic record. Although the visual art is sometimes characterized as symbolic, these representations are actually *iconic*; that is, they have some form relationship to the objects they represent.[31] Musical notes are rarely used as symbols. But Upper Paleolithic sites have yielded some engraved and painted signs that probably refer to things to which they lack any obvious form relationship. They include animal bones engraved with sequences of marks that have been interpreted as mnemonic devices (or "artificial memory systems"). Some of these artifacts may have been used as lunar calendars.[32]

The most symbolic form of external thought—syntactic language—is not directly represented in the archaeological record until the development of writing. And language is almost universally regarded as the centerpiece of modernity. It is the pattern of creativity in the archaeological record of anatomically modern humans after 50,000 years ago that provides the strongest indication of language. The variable structures of Upper Paleolithic art and technology exhibit a complex hierarchical organization with embedded components that resemble the subordinate clauses of phrase structure grammar.

The Phenomenon of Mind

One can hold that the mental is emergent relative to the merely physical without reifying the former. That is, one can maintain that the mind is not a thing composed of lower level things—let alone a thing composed of no things whatever—but a collection of functions and activities of certain neural systems that individual neurons presumably do not possess.

MARIO BUNGE

One of the most important technological breakthroughs in human history occurred in the late thirteenth century. In the years leading up to

1300, someone in Europe—whose identity remains a mystery—created the escapement mechanism for a weight-driven clock. The escapement controls the release of energy (in this case, stored in a suspended weight), which is essential to producing the regular beat of a clock. The perceived need for a timepiece more reliable than the water clock or sundial apparently arose from theology and the requirements of ritual—the obligation to offer prayers at specific times during the day. During the fourteenth century, as clocks spread across western Europe, they acquired secular functions as well.[33]

Clock makers generated a number of major improvements and innovations in the years that followed (for example, the spring drive around 1500) and clock making became the supreme technical art of Europe. The impact of the mechanical clock extended far beyond the making and using of clocks, however, and the historian of technology Lewis Mumford famously designated it the "key machine of the modern industrial age" (figure 1.6) The clock not only influenced other mechanical technologies after the thirteenth century, but became a model and metaphor for the universe—intimately connected to the mechanistic worldview

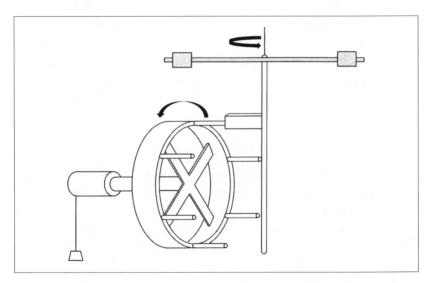

Figure 1.6   The early weight-driven clocks were the digital computers of the sixteenth century and helped establish the mechanistic worldview of the modern age. (Redrawn from Donald Cardwell, *The Norton History of Technology* [New York: Norton, 1995], 39, fig. 2.1)

that emerged in the sixteenth through eighteenth centuries. The Europeans began to think of the universe as a gigantic machine and God as a clock maker.[34]

How did humans fit into this brave new world? René Descartes (1596–1650) thought that the human body—with its various organs circulating the blood, digesting food, and so forth—might indeed be regarded as "a machine made by the hand of God," and he compared it directly with the "moving machines made by human industry." But he made a fundamental distinction between the body and the mind, with its "freedom peculiar to itself." Descartes could not explain the seemingly independent workings of the mind—so clearly manifest in the infinite possibilities of language—as part of a mechanical universe, and he placed it in a separate category.[35] He was reaffirming a view of the mind compatible with both Greek philosophy and Judeo-Christian theology.

Descartes's mind–body dualism set the stage for a debate that continued through the years of the Industrial Revolution to the present day.[36] For the most part, his views have been subject to intense criticism and even ridicule. His contemporary Thomas Hobbes (1588–1679) believed that the mechanistic model could be applied to the mind: "nothing but motions in some part of an organical body."[37] Most of Descartes' critics have been appalled at the implication that supernatural forces are at work in the brain—the "ghost in the machine," wrote Gilbert Ryle (1900–1976). Some philosophers have sidestepped this problem by recasting dualism in terms of properties (rather than substances); they envision the human mind as a material object, but one that operates according to a unique set of principles or properties.

In the mid-nineteenth century, Charles Darwin and Alfred Russel Wallace proposed a "mechanism"—natural selection—to account for the structure and diversity of living organisms and to explain how they have adapted (often in remarkable ways) to the environments they inhabit. Perhaps the most revolutionary aspect of this idea is the recognition that populations of plant and animal species are composed of unique individuals—the variation on which selection acts.[38] Once again, the insights offered by technology played a role in formulating new ideas about how the world works: Darwin devoted the first chapter of *On the Origin of Species* to the practice of plant and animal breeding (that is, artificial selection).[39]

Subsequent discoveries in genetics were necessary to explain how the variation among individual organisms in a population is produced. Darwin died without knowing the underlying genetic mechanisms of the evolutionary process. This did not preclude him, however, from developing a general explanation of how organic evolution works. The later discoveries in genetics were comfortably incorporated into the natural selection model and became part of the neo-Darwinian synthesis of the twentieth century.[40]

Evolutionary biology provided a new framework for understanding how humans fit into the universe, and Darwin was confident that the mind as well as the body could be explained as the product of the evolutionary process. He once characterized thought as "a secretion of the brain." Body and mind alike could be understood as primarily the result of selection for adaptive characteristics in the context of the environments that humans had inhabited over the course of their evolutionary history. Although considered something of a footnote in life-science history, Wallace dissented from this view and endorsed a dualist perspective.[41]

Today, a Darwinian approach to the mind is aggressively pursued by a group of people who describe themselves as evolutionary psychologists. They reflect the marriage of cognitive psychology (which arose from the ashes of behaviorism in the 1950s and 1960s, in the form of the "cognitive revolution") and evolutionary biology. But their "computational theory of the mind" may owe as much to the twentieth-century equivalent of the mechanical clock as it does to biology and psychology theory. The evolutionary psychologists view the mind as a "naturally selected neural computer"[42] that has acquired an increasingly complex set of information-processing functions in response to the demands of navigation, planning, foraging, hunting, toolmaking, communication, and sociality.[43] According to Steven Pinker, art, music, and other creations of the mind that contribute little or nothing to individual fitness are "nonadaptive by-products" of the evolutionary process.[44]

Evolutionary biology provides an essential context for the modern human mind because, despite the curious properties that it eventually acquired (that is, the "freedom peculiar to itself"), the mind is clearly derived from the evolved animal brain. At the same time, an evolutionary context helps identify the point at which the mind departs from the brain and emerges as a unique phenomenon. The story has been pieced together from the fossil record and comparative neurobiology.

The brain evolved more than 500 million years ago among organisms that had become sufficiently complex and mobile that they required information about their environment in order to stay alive and reproduce (that is, continue to evolve). The basic function of the brain or central nervous system is to receive, process, and respond to information concerning changes in the world outside the organism. From the outset, the brain has generated *symbols* because the external conditions (for example, increased salinity or approach of a predator) have to be converted into information; they can be *represented* but not reproduced inside the brain. These symbols are created in electro-chemical form by specialized nerve cells that remain the fundamental unit of the brain.[45]

The brain made complex animal life possible, and its appearance triggered an evolutionary explosion of new taxa that began to lead increasingly complicated lives. Over the course of several hundred million years, some of these taxa—especially the vertebrates—evolved large and complex brains that could receive and process substantial quantities of information from the external world. These brains also could store vast amounts of information (in symbolic form) about the environment, including detailed maps, and they could compute the solutions to simple problems, such as how to climb a particular tree to retrieve an elusive fruit.

From an impressive group of primates, the Old World monkeys and apes, humans inherited a large brain with an exceptional capacity for processing visual information and solving problems. As already described, humans developed the highly unusual ability to externalize information in the brain—probably as an indirect consequence of bipedalism and fore-limb specialization related to toolmaking—and this ultimately became the basis for the mind. As their brains expanded far beyond the already large brain: body ratio of the higher primates, humans evolved the unique capacity to recombine bits of information in the brain in a manner analogous to the recombination of genes (Richard Dawkins suggested that the bits of information could be labeled memes).[46]

The human brain began to *generate* information, rather than merely process it, and to create representations of things that never were but might be. It began to devise not only structures that exhibit characteristics found in living organisms (for example, hierarchical organization), but also structures that are not found in living organisms (for example, geometric design). Moreover, humans had the means—first the hand and later the vocal tract—to reverse the function of the brain: instead of receiving information about the environment, they could imagine something different and create it. Most humans now dwell in a bizarre landscape largely structured by the mind rather than by geomorphology.

The principles that underlie the capacity for creativity—the core property of language and all other media through which humans can imaginatively combine and recombine elements into novel structures—remain an elusive fruit. They have been characterized as emergent properties of the human brain yet to be understood, just as the principles of organic evolution once lay beyond comprehension.[47] There is a consensus of sorts among philosophers, psychologists, and neuroscientists (as well as artificial-intelligence researchers) that these properties emerged from the immense complexity of a brain, which has been described as "the most complicated material object in the known universe." Among psychologists, this view is labeled *connectionism*. Each individual brain contains an estimated 100 billion neurons and a staggering 1 million billion connections (or synapses) among the neurons.[48]

Modern humans added yet another layer of complexity to this most complicated object; by externalizing information in symbolic form with language, they created a *super-brain*. Like a honeybee society, modern humans can share complex representations between two or more individual brains. With language (as well as visual art and other symbolic media), coded information can move from one brain to another in a manner analogous to the transmission of coded information within an individual brain. Each modern human social group, usually defined by a dialect or language, represents a "neocortical Internet" or integrated super-brain. And the archaeological record reveals that it is only with evidence for externalized symbols that traces of creativity, or discrete infinity, emerged roughly 75,000 years ago.

If the principles that underlie creativity continue to elude understanding, other unique aspects of the modern human mind are readily apparent. Because of the human capacity for externalizing thought, the mind transcends not only biological space (the individual brain), but also *biological time* (the life span of the individual). And because, as the psychologist Merlin Donald observed, of the human capacity for externalizing thought in the form of "symbolic storage" (for example, notation and art), information may be stored outside the brain altogether.[49] Indeed, there is no practical limit to the size and complexity of the human mind because it exists—and has existed for 50,000 years or more—outside the evolved brain of the individual organism. The mind comprises an immense mass of information that has been accumulating and developing since the early Upper Paleolithic.

Until recently, the mind was dispersed across the globe in many local units ultimately derived from the initial dispersal of modern humans within and eventually out of Africa about 50,000 years ago.[50] New technologies have begun to effect a global integration, leading back toward what probably was the original state of the mind—a single integrated whole.

Leviathan

Consciousness is, therefore, from the very beginning a social product.
KARL MARX AND FREDERICK ENGELS

The primary focus of Thomas Hobbes's attention was not the mind–body problem, but society and government. In 1651, he published his controversial book *Leviathan, or the Matter, Forme and Power of a Common-wealth Ecclesiasticall and Civill* (figure 1.7). Not long after his death in 1679, Oxford University called for burning copies of it.

Hobbes reversed Descartes's comparison of the living body to a machine and asked: "[W]hy may we not say that all automata (engines that move themselves by springs and wheeles as doth a watch) have an artificiall life?"[51] He promptly extended this notion to human society itself, which he described as "that great Leviathan . . . which is but an Artificiall Man." He likened the various aspects of seventeenth-century European society to the functioning parts of the human body—magistrates as joints, criminal justice system as nerves, counselors as memory, and so forth—joined together by the ability of people to form a collective mind, "a reall unitie of them all, in one and the same person."

By comparing human society to an organism, Hobbes was invoking a parallel with those well-known "Politicall creatures" bees and ants, which are today included among the *eusocial*, or true social, insects. At the beginning of the twentieth century, the entomologist William Morton Wheeler proposed that an ant colony be considered a form of organism, and some years later he introduced the term *super-organism* to describe it. An ant colony, he noted, functions as a unit, experiences a life cycle of growth and reproduction, and is differentiated into reproductive and nonreproductive components.[52] The super-organism concept may also be applied to bees, wasps, and termites. It fell out of favor for some years, but has been revived by Edward O. Wilson and others in recent decades.[53]

Hobbes seems to have been aware of the differences as well as the parallels between human and insect society. While among ants and bees "the Common good differeth not from the Private," in human society, "men are continually in competition for Honour and Dignity" and "there ariseth on that ground Envy and Hatred, and finally Warre." Even worse, while

Figure 1.7 *Leviathan*: The frontispiece of Thomas Hobbes's famous book about society and government, published in 1651, depicts a monarch as being composed of all the individuals in his kingdom, which Hobbes described as "an Artificiall Man." Modern human behavior seems to have emerged with the integration of brains through symbolic language, which allows representations to move from one brain to another analogous to the flow of information in an individual brain.

ants and bees follow the administration program, "amongst men, there are very many, that thinke themselves wiser, and abler to govern the Publique, better than the rest; and these strive to reforme and innovate, one this way, another that way; and thereby bring it into Distraction and Civill warre."[54] In evolutionary biology terms, the differences between eusocial insect and human sociality reflect the contrast between a society based on a super-organism and one based largely on a super-brain.

The key to the insect super-organism is the close genetic relationship among its members. The issue was addressed by Darwin, who—noting that the existence of nonreproductive castes among these species seems to contradict the natural selection model—observed that among social insects, "selection may be applied to the family."[55] The concept was eventually termed *inclusive fitness* and simply recognizes that cooperative behavior will evolve among close relatives when it increases the frequency of shared genes in the population—in these insects, among sibling work-ers.[56] Reproductive biology is thus another important factor among euso-cial animals; they must be capable of producing many offspring to fill the ranks of their societies.[57]

Outside the immediate family, of course, modern human society seems to be based more on the sharing of ideas than of genes. As the evolu-tionary geneticist Richard Lewontin discovered several decades ago, the overwhelming majority of genetic variability (85 percent) lies within rather than between living human populations.[58] The perceived kinship of indi-viduals within races and ethnic groups is largely illusory—these are shared mental representations rather than biological units. Even in hunter-gath-erer societies, where the family and clan are the basic units of organization, marriage practices (that is, the incest taboo) ensure a comparatively low coefficient of relationship.

While honeybees evolved both a super-brain (sharing representations by means of the "honeybee dance") and a super-organism, humans—constrained by their reproductive biology—evolved only the former. The result, it seems to me, is that human societies are inherently dynamic and unstable because the shared thoughts or information that creates the super-brain is constantly disrupted by the competing goals ("Civill warre") of their organic constituents (individuals and/or families).

The modern human super-brain nevertheless seems essential to the exis-tence of the mind. As noted earlier, archaeological evidence suggests that creativity—which has been critical to the growth and development of the mind—was possible only with the confluence of individual brains effected by language and other forms of symbolic communication. And the integrative mechanisms of the super-brain (that is, the externalization of representations in an individual brain) are the means by which the mind exists beyond biological space and time, accumulating non-genetic information (much of it stored outside individual brains) from one generation to the next. Why hasn't the concept been more central to the social sciences? It must be remembered that while copies of *Leviathan* are no longer burned, Hobbes's ideas remain controversial; they seem perfectly compatible with modern totalitarianism. Both the religious and the political doctrines of the modern age that he helped introduce stress the importance of individual choice and action.

Most people are confident, moreover, that creativity is an individual possession, not a collective phenomenon. Despite some notable collaborations in the arts and sciences, the most impressive acts of creative thought—from Archimedes to Jane Austen—appear to have been the products of individuals (and often isolated and eccentric individuals who reject commonly held beliefs). I think that this perception is something of an illusion, however. It cannot be denied that the primary source of novelty lies in the recombination of information within the individual brain, apparently concentrated in the prefrontal cortex of the frontal lobe.[59] But I suspect that as individuals, we would and could accomplish little in the way of creative thinking outside the context of the super-brain. The heads of Archimedes, Jane Austen, and all the other original thinkers who stretch back into the Middle Stone Age in Africa were stuffed with the thoughts of others from early childhood onward, including the ideas of those long dead or unknown. How could they have created without the collective constructions of mathematics, language, and art?

A few years ago, the anthropologist Robin Dunbar suggested that the pronounced expansion of the human brain before the emergence of modernity was tied to human social behavior. Observing a correlation between the size of the neocortex and the size of the social group among various primates, Dunbar concluded that the enlarged *Homo* brain is a consequence of widening social networks and that language evolved as a means to maintain exceptionally large networks.[60] I would suggest a slightly different version of the "social brain hypothesis." Modern human sociality is based primarily on the sharing of information—massive quantities of information—which requires considerable memory-storage capacity.

I think that humans evolved oversize brains when they began to share very large amounts of information.

The super-brain and the collective mind might explain fundamental aspects of modern human cognition, including what philosophers refer to as *intentionality*—the ability to think about objects and events outside the immediate experience of an individual—as well as the "sense of self" and even consciousness. I think it is possible that all these phenomena are consequences of a collective mind (even if it has no collective sense of self or consciousness). If we conceive of the super-brain as a cognitive hall of mirrors that permits each individual to reflect the thoughts of others, perhaps it provides the basis for mutual recognition of the thinking individual. Perhaps it is a biological imperative for the constituents of a super-brain—who are *not* components of a super-organism—to firmly distinguish themselves as individual thinkers within a collective mind.

In recent years, partly as a result of advances in brain-imaging technology, neuroscientists like Gerald Edelman have begun to invade areas once left to philosophers. Edelman draws a fundamental distinction between primary consciousness and higher-order consciousness, and he suggests that the more evolved metazoans like birds and mammals possess *primary consciousness*: "the ability to generate a mental scene in which a large amount of diverse information is integrated for the purpose of directing present or immediate behavior."[61] Animals with primary consciousness appear to lack a sense of self or a concept of the past or the future. In fact, they seem to lack any ability to think in conceptual or abstract terms.

Only modern humans possess what Edelman refers to as *higher-order consciousness*, which entails an awareness of self, abstract thinking, and the power to travel mentally to other times and places. This apparently unique form of consciousness is possible only with "semantic or symbolic capabilities." In order to think about objects and events that are removed from the immediate environment of an organism, symbols shared among multiple brains like words and sentences are necessary. Thus Edelman links higher-order consciousness to language. But the uniquely human form of consciousness is also inextricably tied to social interactions because language is a social phenomenon; the meanings of words and sentences are established by convention (and become a component of the collective mind).[62]

Philosophie der Technik

> Technology is a way of revealing. If we give heed to this, then another
> whole realm for the essence of technology will open itself up to us. It
> is the realm of revealing, i.e., of truth.
>
>   MARTIN HEIDEGGER

The earliest formal use of the term "philosophy of technology" is ascribed to
a rather obscure German-Texan named Ernst Kapp (1808–1896). Kapp was
a "left-wing Hegelian" who sought to develop a materialist version of Hegelian philosophy like his more famous contemporary Karl Marx. Charged
with sedition, he fled Germany in the late 1840s and settled in central Texas.
Kapp eventually returned to Germany and, in 1877, published *Grundlinien
einer Philosophie der Technik*, which outlines a philosophy of technology—
apparently reflecting his experience in the Texas wilderness—based on the
notion that tools and weapons are forms of "organ projection."[63]

Although the influence of his ideas was limited, Kapp initiated a philosophical tradition in Germany that continued into the twentieth century.
It drew on the thinking of major figures in German philosophy, especially
Immanuel Kant. Frederick Dessauer (1881–1963) suggested that technology provides the means to engage what Kant had defined as "things-in-
themselves" (*noumena*) as opposed to mere "things-as-they-appear" (*phenomena*). Dessauer perceived that only the craftsman and inventor encounter
noumena, not the factory assembly-line worker and the consumer of mass-
produced technologies.[64] Like Kapp, Dessauer had to flee Germany for
political reasons—in his case, after the Nazis came to power in 1933.

Ironically, perhaps, the clearest philosophical statement on technology
may have come from Martin Heidegger (1889–1976), who was both a neo-
Kantian and a member of the Nazi Party until 1945. Several years after
the end of the war, Heidegger gave a series of lectures in Bremen that
included thoughts on *philosophie der technik*. He was sharply critical of what
he referred to as the "instrumental" view—that technology is merely "a
means and a human activity" analogous to the use of natural objects by
some animals. Instead he suggested, apparently building on both Kant and
Dessauer, that technology is a "way of revealing" the world—of acquiring
knowledge and learning the "truth" (*Wahrheit*). Technology "makes the
demand on us to think in another way."[65]

Why wasn't technology more widely and fully addressed by philosophers in western Europe or other parts of the industrialized world? Given

the pivotal role of technology in transforming European society and thought after 1200, the avoidance of a philosophy of technology outside Germany seems more than odd; it borders on the pathological. During the twentieth century, one factor was an acute awareness of the negative effects of many new technologies; between 1914 and 1945, the most striking practical applications of chemistry and physics seemed to be high-tech murder.

There was a tradition with deeper roots in the Industrial Revolution, expressed primarily through literature and later through film, of regarding technology as a separate and potentially threatening entity. The theme was woven into Mary Shelley's *Frankenstein* in the early nineteenth century and taken up a few years later—in response to Darwin—by the novelist Samuel Butler, who warned that machines were evolving faster than plants and animals and subordinating humans to their needs.[66] The dangers of runaway technology have remained a steady theme in Western popular culture.

But technology is an essential part of the mind, and a philosophy of technology must be incorporated into any theory of mind. Technology stands in relation to the mind—an integrated and hierarchically structured mass of information—as the phenotype of an organism stands in relation to its genotype. Technology is the means by which the mind engages the external world, as opposed to language and art, which are the means by which parts of the mind communicate with each other. The distinction reflects the observation that the world external to the brain is not the same as the world external to the mind, since much of the mind already exists outside the brains of individual humans.

Comparative ethology suggests that a lengthy history of animal "technology" probably exists among many lineages. It includes the use and modification of natural features and objects by a diverse array of complex animals—for example, the excavation of intricate burrow systems and the construction of nests from many parts. Humans inherited a particularly impressive pattern of tool use from the highly manipulative primates. As Jane Goodall revealed several decades ago, this pattern includes the making of simple tools, such as termite-fishing implements, by chimpanzees in the wild.[67]

The emergence of the human mind nevertheless transformed technology into a very different phenomenon. With their talent for externalizing mental representations by their hands—first evident among some African hominins by 1.7 million years ago—humans acquired the ability to translate their thoughts into phenotypic form.[68] As noted earlier, they reversed the function of sensory perception by converting the electro-chemical symbols in the brain into objects that had no prior existence in the landscape. At first, the structure of these objects was simple, but larger-brained humans who evolved roughly 500,000 years ago began to make more complex, hierarchically structured implements. The creative potential of the modern human mind, evident in the archaeological record after 100,000 years ago, removed all practical limits to the complexity of technology.

From the outset, this new form of technology was a collective enterprise. If each Lower Paleolithic biface was unique, like each individual within a population, the mental template of each biface was the same (or very similar) and clearly was shared among many individual brains in space and time. The emergence of symbolic communication in the form of syntactic language created a super-brain and thus vastly increased, it seems, both the computational and the creative powers of human thought by pooling the intellectual resources of each social group.

The unprecedented size of a few middle Upper Paleolithic settlements indicates that the number of constituent brains within social groups in some places, even if assembled only temporarily, had increased significantly by 25,000 years ago. These groups possessed a larger and presumably more powerful super-brain. More permanent-looking settlements of comparable and even greater size appeared after the Last Glacial Maximum, and some of them anticipated the sedentary villages and farming communities of the postglacial epoch.

The number of individuals in residential farming communities, which experienced explosive growth in some places after 8,000 years ago, triggered a radical reorganization of society and economy along rigid hierarchical lines. The mind of the early civilizations reflected this hierarchical socioeconomic structure. Two trends already evident during the preceding period were greatly expanded: the specialization of technological knowledge and the (hierarchical) organization of collective undertakings in various technologies (for example, irrigation systems and public buildings). New technologies facilitated the communication, manipulation, and storage of information.

The size and organization of the civilized collective mind produced technology on a vastly greater scale of complexity, although the rigid hierarchies of the early civilizations seem to have constrained creativity and suppressed its potential for many centuries. West European societies somehow loosened these constraints after 1200 and began to create an industrial civilization. The complex technologies of late industrial and post-industrial societies reflect the enormous growth of specialized knowledge and the hierarchical organization of technology. Robert Pool described the construction of a complex piece of technology in the modern era: "No single person can comprehend the entire workings of, say, a Boeing 747. Not its pilot, not its maintenance chief, not any of the thousands of engineers who worked on its design. The aircraft contains six million individual parts assembled into hundreds of components and systems. . . . Understanding how and why just one subassembly works demands years of study."[69]

The immense growth of human populations and the rise of complex organizations in the postglacial epoch were consequences of the steady accumulation of knowledge about the world—knowledge about how to manipulate and redesign the landscape to suit human needs and desires. And most of this knowledge was derived from what Heidegger referred to as the "revealing" power of technology. It was not accumulated by individual humans or even by the super-brain, but was incorporated into the collective mind from one generation to the next.

Today most humans inhabit an environment that bears little resemblance to the naturally evolved landscapes their ancestors occupied a million years ago. The processes of neither Earth history nor evolutionary biology yield the geometric patterns of cities, suburbs, and rural agricultural landscapes that now cover much of the land surface of the planet. These patterns are generated in the mind and projected onto the external world. When human societies were uniformly small and the ability to manipulate the world was limited, their impact was modest. But as populations grew and the body of collective external thought accumulated and expanded, entire landscapes were transformed—one of the most striking consequences of human evolution.

The roots of the mathematical patterns of thought lie deep in the evolutionary past of the brain. As the metazoan brain evolved for information processing and directing the responses of other organs, including navigation, selection clearly favored efficiency (for example, calculating and executing a straight-line path to or from an object or another animal). Thinking evolved along mathematical lines that are not characteristic of organic design, which may often exhibit symmetry, but not fundamental geometric patterns such as triangles and squares. When humans began to translate thought into material objects and features, they imposed new forms on the landscape. Many of these forms reflected the geometry of the mind.

More than a million years ago, humans apparently were externalizing thought only in the form of reshaped pieces of stone and (presumably) wood. The hand axes they produced exhibit a symmetrical oval design in three dimensions that does not necessarily reflect a geometric pattern of thought, and the "spheres" that date to this period seem to represent uniformly battered fragments of rock.[70] During the later phases of the Middle Stone Age in Africa (after 100,000 years ago), simple geometric designs in two dimensions were engraved on mineral pigment, and similar designs appear on both engraved and painted surfaces during the early Upper Paleolithic in Eurasia (after 50,000 to 40,000 years ago). Occupation features that might reflect the imposition of simple geometric patterns on the ground include linear arrangements of hearths and circular pits in middle Upper Paleolithic sites (around 25,000 years ago).

It is not, however, until the late Upper Paleolithic (after 18,000 years ago) that unambiguous examples of geometric design appear on the landscape, in the form of square and rectilinear house floors. The art of the period is also rich in paintings, engravings, and carvings of various geometric patterns. In the early postglacial epoch, villages comprising an amalgamation of rectilinear houses were constructed in the Near East, and they eventually were superseded by towns and urban centers with street systems and public monuments and plazas. Villages, towns, and cities were surrounded by agricultural landscapes composed of orthogonal irrigation networks and linear crop rows.[71]

Most of the human population now resides in a bizarre environment structured by the collective mind. Rock, soil, and water bodies have been altered, and naturally evolved ecosystems have been re-created as lawns, gardens, and planted fields. Many humans spend the entire day within the confines of rectilinear walls, stairways, doors, and sidewalks. Moreover, humans born into this landscape of thought internalize it as they mature, and the process is reversed: the external structure is incorporated into the synaptic pathways of the developing mind.

A Fossil Record of Thought

The archaeological record is constituted of the fossilized results of human behavior, and it is the archaeologist's business to reconstitute that behavior as far as he can and so to recapture the thoughts that behavior expressed. In so far as he can do that, he becomes a historian.

V. GORDON CHILDE

In the early twenty-first century, there is no field of research more exciting than neuroscience. This is largely a consequence of brain-imaging technologies developed in the 1970s and later, which provide measures of neural activity. The new technologies include computed tomography (CT), positron-emission tomography (PET), and functional magnetic resonance imaging (fMRI). For the first time, researchers can observe brain function—including proxy measures of thought itself—in a manner analogous to the examination of physiological functions like blood circulation and digestion. The increasingly broad application of the new techniques has produced many revelations and surprises,[72] and some neuroscientists have begun to address questions once left to philosophers of the mind.

If current developments in neuroscience offer unprecedented insights into the brain and the mind, the field of archaeology would seem to be one of the least promising avenues. Archaeology began as a hobby for collectors of antiquities and emerged during the nineteenth century as a discipline devoted to the documentation of material progress. Human prehistory was initially subdivided into three successive ages: Stone, Bronze, and Iron. The emphasis on progress reflected a European perspective on history and humanity that dominated but did not survive the nineteenth century.[73]

V. Gordon Childe is often quoted for his observation that the material objects of the archaeological record must be treated "always and exclusively as concrete expressions and embodiments of human thoughts and ideas."[74] Archaeologists have tended to treat them, however, more along the lines of what Martin Heidegger labeled "instruments."[75] The artifacts and features of the archaeological record typically have been viewed as a means to an end—survival and progressive improvement—somewhat analogous to the computational theory of mind. The instrumental approach was on the march in the years following World War II, as many archaeologists adopted a *systems ecology* perspective.[76]

In the final decades of the twentieth century, a reaction arose against this approach (often condemned with the ultimate epithet "functionalism"), and the views of Childe were recalled.[77] In the early 1980s, the British archaeologist Colin Renfrew urged his colleagues to develop an archaeology of mind, or cognitive archaeology.[78] In 1991, the psychologist Merlin Donald published a book about the origin of the mind in which he observed that modern humans are storing thoughts outside the brain.[79] This idea made its way into paleoanthropology and stimulated further reflection and debate among archaeologists. Renfrew and other cognitive archaeologists began to think about the way the mind engages the material world through the body.[80]

How does the cognitive archaeologist go about collecting and analyzing "human thoughts and ideas" from the archaeological record? If written records or accessible oral traditions are present, they provide an interpretive context for the archaeological materials. In most prehistoric settings, written and oral sources are absent, but other types of symbols are—at least after 75,000 years ago. If they are iconic (for example, a female figurine), their referents are knowable, if not their possible wider meaning (perhaps fertility). But if they are abstract symbols with an arbitrary relationship to their referents, their encoded meanings may be unknowable. And before 75,000 years ago, symbols seem to be absent altogether from the archaeological record.

Instead of trying to recover content and meaning, the cognitive archaeologist can focus his or her attention on the *structure* of externalized thoughts. This opens a wider and deeper field of inquiry because humans have been externalizing their thoughts in nonsymbolic form for more than 1.5 million years, creating a *fossil record of thought* that is unique in the history of life. How are these representations organized? What is the structural relationship among the elements—digital and/or analogical—from which they are composed? How many hierarchical levels and embedded components can be identified? How much creative recombination of elements is manifest among different artifacts? Moreover, the symbolic representations that emerge after 75,000 years ago exhibit their own structure.[81]

In fact, the formal analysis of structure had its origin in the study of language (that is, symbolic representation). It was initially developed by Swiss-born linguist Ferdinand de Saussure (1857–1913), who taught in Paris during the early years of the twentieth century. *Structuralisme* was applied in other fields by various French scholars, including the ethnologist Claude Lévi-Strauss (1908–2009), who analyzed the structure of kinship, myth, song, and other aspects of "primitive" culture.[82] André Leroi-Gourhan (1911–1986) developed a structuralist approach to Paleolithic archaeology; not surprisingly, perhaps, he saw parallels between language and toolmaking. Although he became famous for his study of cave art, his most significant contribution was the *chaîne opératoire*, or the sequence of steps involved in making an artifact.[83] Structuralism later faded in popularity, but the sequential analysis of artifact production became a core element of archaeology. It is an important source of information about the increasingly hierarchical structure of artificial representation in human evolution.[84]

Much of this book is devoted to the *modern mind*, which emerged before 50,000 years ago in sub-Saharan Africa and spread across the Earth with anatomically modern humans (*Homo sapiens*). The advent of the modern mind is discernible only through the fossil record of thought (the modern human cranium had earlier acquired its volume and shape). As already noted, the archaeological record reveals the potentially infinite capacity for creative recombination of elements in various media among *Homo sapiens* after 50,000 years ago and at a level of structural complexity commensurate with that of living people. And this evidence coincides with archaeological traces of symbolism—the sharing of complex mental representations through icons and digital symbols in the form of art and, by implication, spoken language—which was a prerequisite for the super-brain and collective mind. In fact, there are now indications of the collective mind at least 75,000 years ago in the form of abstract designs incised into mineral pigment. But broader spectra of space and time are still required to document creativity on a modern scale (that is, the later Middle Stone Age of Africa and the early Upper Paleolithic of Eurasia and Australia).

With the emergence and dispersal of the modern mind, the process of accumulating knowledge in the form of information shared among individual brains and stored in many forms of external thought began in earnest. Much of this knowledge was acquired through the process of technological engagement with the external world (or with "things-in-themselves," as Kant would say). The information was stored, communicated, and manipulated in many different forms (language, visual art, music, tools, and so on), and at least some of these media were preservable materials. Most of the subsequent growth and development of the modern mind occurred in the absence of written records. Nevertheless, the accumulating mass of information and knowledge is partially observable in the fossil record of thought—the archaeological record.

As human populations in some places experienced explosive growth after the Upper Paleolithic—a consequence of their accumulated knowledge, as well as favorable climate conditions—the mind became immense. Not only were hundreds of thousands of individual brains contributing to the collective mind in places like Mesopotamia and northern China, but the storage of information had reached vast proportions. These societies reorganized themselves, often with violence or threats of violence, along rigid hierarchical political and economic lines. The collective mind also was reorganized into a more hierarchically structured entity with specialized compartments of thought (for example, metallurgy and architecture). Writing was one of several means by which the sprawling mind of civilization was integrated and organized, and it eventually broadened the record of thought in profound ways.

This book also is about the origin of the mind, or what the archaeologist Steven Mithen described as the "prehistory of the mind."[85] The fossil record of thought provides critical data here as well, although in necessary conjunction with the other fossil record: the skeletal remains of earlier humans. This is because developments in human evolutionary biology—including changes in the brain, hand, and vocal tract—over several million years are implicated in the origin of the mind. The fossil record of thought offers a glimpse of the externalized mental representations of early humans, which not only are less structurally complex than our own, but also exhibit little creative variation; most of the latter appears to be random and aimless. I suspect that these early specimens of external thought played a role in the origin of mind and that they are more than a fossil record of early thought or products of what might be referred to as the *proto-mind*.

The emergence of the mind probably is connected to external thought; the projection of mental representations outside the brain by means of the hand and, later, the vocal tract is related to the expansion of the brain and the development of its unique properties of creativity and consciousness. As a result of bipedalism, the early human brain began to engage directly with the material world not simply by moving the body, but also by externalizing mental representations—imposing their nongenetic structure on the environment with the hand—and thus establishing a feedback relationship between the brain and the external thoughts, as well as among the multiple brains within each human social group. Making an artifact is a way of talking to oneself. As the brain expanded and the synaptic connections grew exponentially, the externalized thoughts became increasingly complex and hierarchical, as well as more creative.

Once humans began to shape natural materials in accordance with mental templates (around 1.7 to 1.5 million years ago), they began to populate their environment with physical objects that reflected the structure of mental representations (specifically, artificial representations). Semantic thoughts acquired a presence outside the brain, and they were visible not only to the individuals who made the artifacts but also to others; they became perceptual representations of unprecedented form (that is, the brain receiving representations of its own creation). The artifacts were talking back. These circumstances created a reciprocal relationship between internal and external thought, as well as a dynamic interaction among individuals projecting and receiving artificial representations.

As the thoughts externalized by humans became increasingly complex (beginning after 500,000 to 300,000 years ago), the synaptic pathways of the brain, where information is stored, contained a growing percentage of received representations generated by other human brains. Before the emergence of language, representations would have been communicated among individuals and across generations in the form of artifacts. With their hands, humans could manipulate and recombine the elements of these external representations. And at some point (perhaps not until after 500,000 years ago), external representations in the form of more efficient implements were almost certainly having a measurable impact on human ecology and individual fitness—another reinforcing feedback loop (that is, natural selection).

The critical transformation from the proto-mind to the mind probably began roughly 300,000 years ago, when brain volume had attained its modern level and the vocal tract was either evolving or had already evolved its modern human form. Changes in the shape and size of the brain probably were important, including relative expansion of the prefrontal cortex and temporal lobes. The archaeological evidence, which includes composite tools and weapons, as well as a variety of small biface forms, indicates the ability to externalize representations in a more complex and hierarchical form with greater variation in design (that is, more creative).[86]

Among unanswered questions about human evolution, the origin of language is still at the top of the list. In some respects, it is more of a mystery than consciousness itself. But if the immediate causal factors that lie behind the emergence of spoken language remain obscure (perhaps a genetic mutation), the ultimate source most probably is the increasingly creative manipulation of materials with the hands—the ability to recombine the elements of artifacts on multiple hierarchical levels. The same cognitive faculties apparently underlie language and toolmaking. And at some point (perhaps about 100,000 years ago or earlier in Africa), the vocal tract acquired a specialized motor function analogous to that of the hands.

Archaeologists have long speculated about the parallels and possible evolutionary relationship between creating artifacts and constructing sentences.[87] The subject received a good deal of attention during the 1960s and 1970s, beginning with Leroi-Gourhan. In 1969, Ralph Holloway—a specialist on the evolution of the human brain—wrote a classic paper on the underlying similarities between language and toolmaking.[88] Others who articulated this view included the American archaeologist James Deetz and the South African paleoanthropologist Glynn Isaac.[89] In the decades

that followed, the parallels were pursued with less vigor, owing perhaps
to the rise of the "modular model" of the mind, which implies separate
neural structures for each language and toolmaking.[90] But in a major paper
published in 1991, the psychologist Patricia Greenfield restated the case
for a "common neural substrate" for language and "object combination."[91]

Brain-imaging research during the past two decades has eroded support
for the modular model and yielded new information on possible neural
connections between language and tools. In general, brain-imaging studies
have revealed a high degree of interconnectivity among the various regions
of the brain.[92] Some recent PET data indicate activation of many of the
same areas for spoken and gestural language, supporting earlier views of
neural connections between vocal tract and hand function.[93]

The turning point—the emergence of the modern mind—was the combined appearance of syntactic language, unlimited creativity, and the
super-brain. The fossil record of thought seems to indicate that these developments were more or less coincident, and they are logically interrelated,
if not inseparable. Syntactic language created the super-brain; earlier forms
of shared thought would seem to have been too limited. At the same time,
the creative manipulation of information that language represents was a
social phenomenon that could exist only among multiple brains in biological space and time. And the exponential growth in neuronal network
complexity engendered by the integration of brains, already oversize, in a
human social group may have been the requisite basis for unlimited creativity expressed in language and other media.

The seeming suddenness of these developments has inspired the suggestion among some paleoanthropologists that a genetic mutation may
have provided the trigger. Possible mutations have been hypothesized in
the areas of speech and working memory.[94] Moreover, there have been
attempts to research this problem with genetic data from both living
and fossil humans.[95] It turns out that there is yet another fossil record—
composed of DNA rather than mineralized bones or thought—that holds
potential for understanding the origin of the mind.

2

DAYDREAMS OF THE LOWER PALEOLITHIC

I dream things that never were and say why not?

GEORGE BERNARD SHAW

THE EARLIEST KNOWN EXAMPLES of external thought appear roughly 1.5 million years ago in the form of chipped stone objects. They are pieces of rock chipped on both sides into an oval shape that archaeologists refer to as *bifaces*, or bifacially flaked tools. They have achieved the exalted status of an externalized mental representation because they bear little resemblance to the natural object—a variously shaped cobble or fragment of rock—from which they were made. They have been transformed by a pair of human hands in accordance with a mental template, or an internal mental representation.[1]

Archaeologists have been unable to determine whether the bifaces were meant to "stand for something" or were an arbitrary form imposed on pieces of rock. They may have represented a large leaf of exaggerated thickness or a giant coarse seed.[2] And it is possible that the biface makers imposed representations on other natural forms—carving pieces of wood or drawing patterns in sand, for example—that have not been preserved (some rare examples of bifaces chipped in bone have turned up).[3] In any case, humans continued to make bifaces for more than a million years before they began to create more complex representations outside their heads. Many of the later bifaces exhibit a more refined appearance, but their essential form remains unchanged.

The appearance of the bifaces was a major event in the prehistory of the mind, signifying the emergence of what I have termed the proto-mind. From that moment forward, humans began to match at least some of the thoughts generated in their heads with those constructed outside the brain. Mental representations were no longer entirely confined to the brain. Equally important, it seems, was the fact that mental representations—not simply emotional states or simple bits of information—could be communicated from one individual brain to another. Humans now found themselves in a landscape populated not only by physical features and organic beings, but also by thoughts in an incredibly literal sense.

The Evolutionary Biology of the Mind

[M]an bears in his bodily structure clear traces of descent from some lower form; but it may be urged that, as man differs so greatly in his mental power from all other animals, there must be some error in this conclusion.

CHARLES DARWIN

One of the more intriguing implications of the archaeological record is that the mind or something very much like it could have emerged in another form of life. There is nothing magical about humans; it is simply that the evolutionary biology of humankind created certain conditions from which the emergence of the mind was possible and perhaps even inevitable. What conditions would be necessary to produce a true mind, a mass of structured information that exists both inside and outside an integrated group of organically evolved individual brains that continually feed it with newly generated structures of information?

An organism must possess a large brain capable of constructing complex representations from incoming sensory data. Many vertebrates meet this criterion, and some of them have brains that exceed the volume of the human brain either in absolute terms or in relation to their overall body size.[4] And while a brain might construct internal representations of the outside world from a sense of smell or sound, representations based on electromagnetic radiation within the spectrum of visible light would seem to offer the greatest potential for complex and intricate maps of the external world.[5] Vision provides a potentially immense range—for all practical purposes unlimited, when humans look at distant stars and galaxies. Finally, there must be a means to project mental structures outside the brain in order to establish a feedback relationship between internal and external representations. In humans, the hands and later the vocal tract became the means, but any number of features might have evolved an analogous function in other organisms, such as a tail or antennae.

Humans evolved within the primates, and both comparative anatomy and the fossil record indicate that the aforementioned conditions have deep roots in this mammalian order. As early as 50 to 40 million years ago, the primates exhibited three important trends that provided a basis for the subsequent evolution of monkeys, apes, and humans: (1) increased size of the brain relative to the body, (2) increased emphasis on vision and corresponding reduction in sense of smell, and (3) increased use of extremities for manipulation.[6] All were essential for the emergence of the human mind.

The basic function of a brain is to enable an organism to respond to spatial and/or temporal changes in its environment.[7] A brain allows the organism to detect a drop in temperature, a potential food item, the approach of a predator, and other changes important to its survival and reproduction, and to generate an appropriate response. Among all but the most primitive metazoans (that is, sponges), the function of the brain is performed by specialized cells called *neurons*; among most multicellular animals, neurons are organized into a central nervous system. But even the simplest protozoa that appeared more than 3 billion years ago—the prokaryotes, or unicellular organisms that lack a nucleus—contain structures that evolved to perform the same basic function. The notorious and well-studied bacterium *E. coli* has protein molecules embedded in its cell wall that act as receptors to detect specific chemicals outside itself. These receptors trigger responses that vary according to whether the chemical detected is a food or a toxin. Propelled by six flagella, *E. coli* can maintain or change direction in response to the information received and transmitted by its receptors.[8]

When the human mind is placed into this universal functional perspective on the nervous system and brain, its unique capabilities become less curious. Although the ability of the mind to create alternative realities seems rather exotic in the context of metazoan brain evolution, it still serves the same basic purpose. Modern humans not only sense changes in the environment, but imagine and predict future changes. This feat is achieved by generating novel and often complex representations of the environment, either inside or outside the brain and often collectively among individuals. Cognitive psychologists often refer to it as "mental time travel."[9] In fact, the effort to predict animal movements, rains and floods, solar and lunar eclipses, and other natural events has been a major theme in human history.

Neurons and the nervous system evolved more than 500 million years ago among the ancient phylum Cnidaria, which includes jellyfish, hydra, sea anemones, and corals.[10] The design of nerve cells has not changed fundamentally since the appearance of the Cnidarians, only the ways in which they are organized. Neurons are true cells (eukaryotes) with a nucleus, but they also possess a long string-like structure called an *axon* and fine branching structures called *dendrites* (figure 2.1). Because they are dynamically polarized, neurons can transmit electrical signals to each other across *synapses*, the point at which they connect. They can transmit signals in digital (on or off) or analogical (continuous scale) mode. Networked together by synaptic connections, neurons can store, as well as transmit, information about the environment.

Just as there is continuity between the "brain" functions of the simplest organisms and the often wild imaginings of the human mind, responding to changes in the external world, there is continuity between the transmission and storage of information in the neurons of the early metazoans and human language. As the linguist Derek Bickerton observed, both are representational systems.[11] Even the simple nervous systems of the Cnidarians contain information about the environment outside the organisms. As new metazoan groups evolved, they developed more complex neuronal networks to process and store more detailed information—more detailed and accurate maps—of the external world. They also evolved new organs that expanded the amount and type of information received—the eye is

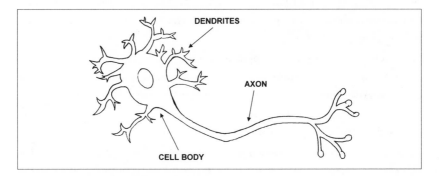

Figure 2.1 The fundamental design of the specialized nerve cell, or neuron, has not changed for 500 million years.

especially important to the evolution of humans—and those that expanded the range of responses to the environment from the organism.

The evolution of the neuron facilitated the great explosion of complex animal life that took place in the early Cambrian period roughly 540 to 520 million years ago. New forms included the worms, which exhibit a bilateral body plan (unlike the radial Cnidarians), with a defined head versus tail as well as dorsal and ventral sides. The first central nervous systems, comprising a brain in the head region and a nerve cord extending to the tail, developed in these phyla (Platyhelminthes and Annelida). The pattern of dorsal brain and ventral nerve cord was repeated in other phyla that emerged during the Cambrian explosion, including mollusks, crustaceans, and insects.[12]

Among these new phyla were the chordates (Chordata), which are distinguished by the embryonic development of a notochord. The first chordates were invertebrate marine organisms, but they subsequently gave rise to marine and terrestrial vertebrates: reptiles, dinosaurs, birds, and mammals with large complex brains. The living *Amphioxus* is thought to be representative of the earliest chordates. It possesses structures that correspond to several major components of the vertebrate forebrain, including the frontal eyes and hypothalamus, as well as the hindbrain; both the nerve cord and the brain are dorsal. The frontal "eye spot" in *Amphioxus* contains pigmented cells that seem to anticipate the vertebrate eye.[13]

Like the mind, the evolutionary process is creative. Although its products are the result of natural selection acting on randomly generated variation, novel forms are often characterized as "innovations."[14] The evolution of the early vertebrates, which began about 470 million years ago, following the end of the Cambrian, entailed a number of such innovations critical to the development of large brains. One of them was the *neural crest*, which eventually gave rise to the cranium, jaws, and teeth. Another was the protein *hemoglobin*, which plays an important role in meeting the high oxygen needs of the brain. Yet another was *myelin*, sheets of protein and fat molecules that insulate the axon and improve its conductivity. Insulated neurons can be packed into a small area to form a dense network.[15]

The early vertebrates evolved frontal eyes and nostrils, as well as several related brain components: the telencephalon and midbrain (mesencephalon) and the cerebellum (figure 2.2). The telencephalon contains expanded olfactory receptors, reflecting heavy reliance on the sense of smell. Of great significance for later brain evolution was the emergence of the midbrain, with *optic tectum* for receiving visual information and integrating data from

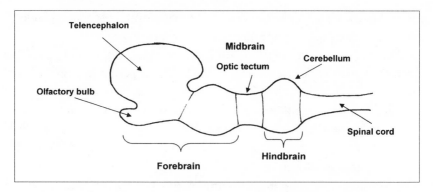

Figure 2.2 The basic vertebrate brain evolved more than 400 million years ago. A significant development was the appearance of the midbrain, with the optic tectum for receiving visual sensory input and integrating information from other senses.

other senses. The cerebellum also plays a role in vision by stabilizing visual representation as the head moves.[16] Successful early vertebrates included the sharks, whose comparatively large brains probably are related to the substantial information needs of a large predator. The first terrestrial vertebrates (amphibians) emerged from the sea about 350 million years ago, and reptiles evolved roughly 300 million years ago.[17]

Disaster struck 50 million years later, at the end of the Paleozoic era, with the most massive wave of extinctions in Earth history, probably a result of cooler climates. Ninety percent of the known fossil taxa vanished. However, the ecological vacuum opened opportunities for the evolution and proliferation of novel life-forms. On land, much of this vacuum was filled by warm-blooded vertebrates, including dinosaurs, birds, and—by 220 million years ago—mammals. The enormous energy demands of maintaining a constant body temperature seem to have been a principal factor in the evolution of bigger brains. Warm-blooded vertebrates need a large amount of accurate and detailed information about their environment.[18]

Among the mammals, this need was addressed in part by an evolutionary innovation in brain structure that was critical to the subsequent emergence of the mind. The *neocortex*—a uniquely mammalian feature—is a six-layer, sheet-like structure that covers the roof of the forebrain. Each layer is connected to other parts of the brain. The neocortex receives, integrates, and stores information about the environment from the various senses. Another innovation was a chain of ossicles that conducts sound from the eardrum to the inner ear, allowing mammals to hear high frequencies.[19]

The brain had acquired a new function especially evident in mammals. It was no longer simply identifying changes in the environment and coordinating appropriate responses to them. The information acquired through the senses and stored in the neuronal networks had become an essential part of how organisms survived and reproduced; learning and memory were now playing a role analogous to those of the genetically controlled (and naturally selected) structures of the body. Unlike fish and reptiles, mammals (as well as birds) require extensive parental care during the early phases of life, which reflects the increased importance of information.

The Visual Animal

Most primate-typical features involve the visual system.
GEORG F. STRIEDTER

The primates are a group of placental mammals that evolved from insectivores toward the close of the Mesozoic era, about 75 million years ago, and apparently occupied arboreal niches in what is now North America and Europe. They diversified after another massive wave of extinctions wiped out the dinosaurs in the late Cretaceous period. The early primates had small brains, long snouts, and laterally placed eyes, but forms resembling the living prosimians—tarsiers, lemurs, and lorises—were present by roughly 50 million years ago (Eocene epoch). They possessed larger brains with smaller olfactory bulbs (indicating reduced reliance on smell), more-forward-facing eyes, and long digits with nails instead of claws. These trends continued among the early catarrhines—a group that includes Old World monkeys and apes—during the Oligocene epoch, which began about 35 million years ago.[20]

Evolutionary innovations in brain structure, especially related to vision, in the Eocene primates took place among the catarrhines and were critical to the emergence of the mind. The neocortex expanded in size and in proportion to the brain as a whole. In the small prosimian *Galago demidovii*, the neocortex comprises 46 percent of brain volume; among living chimpanzees, the neocortex occupies no less than 76 percent of the brain.[21] This surely reflected integration and storage of more information about the environment, but perhaps specifically more information about the social environment. Among primates, the ratio of the neocortex to total brain volume correlates significantly with the size of the social group.[22]

The primates evolved a complex of interrelated changes pertaining to vision, visual representation, and the visual coordination of muscle activity. The density of photoreceptors in the retina increased. The visual cortex of the brain expanded significantly. The optic tectum of the midbrain was restructured in relation to the overlapping fields of vision created by front-facing eyes. It functioned in a more specialized manner to allow primate eyes to fixate on objects of interest. In the forebrain, new components of the visual system improved the perception of motion and of the form of objects. Perhaps most important of all with reference to the subsequent evolution of the human hand was the development of a cortical area, unknown in other mammals, for the visual coordination of motor tasks.[23]

The catarrhines "reinvented" color vision. Most mammals possess limited color vision based on two types of cone photoreceptors. One of these cone types is sensitive to shorter wavelengths of light (blue portion of spectrum), while the other is sensitive to longer wavelengths (red portion of spectrum). As a result of a gene duplication, the ancestor of living catarrhines evolved a third cone type, which had been present among earlier mammals. Human vision is catarrhine vision.[24]

More than any other feature or complex of features, it seems, the visual system distinguishes the primates from other mammals. Why did it evolve so significantly in this group? The forward-facing eyes and overlapping fields of vision have been interpreted as part of an arboreal adaptation.[25] They also have been ascribed to predation on small, fast-moving vertebrates and insects.[26] Perhaps both explanations are valid. The (re)appearance of color vision among the catarrhines has been tied to their reliance on fruit and the need to identify it against a backdrop of green foliage.[27] The rather generalized adaptations of the higher primates may have placed an added premium on a visual system that provides so much information about the environment.[28] But another contributing factor seems likely to have been the growing importance of visual cues (and corresponding decline of chemical signals) in primate society, especially facial expressions.[29]

A theory of human vision was articulated by David Marr in 1982 and remains an important contribution to neuroscience.[30] Marr placed primary emphasis on vision as *information processing* and on the creation of visual representations in the brain.[31] He characterized the process as hierarchical and recursive, and the psychologist Michael Corballis later noted the parallels between Marr's theory of vision and language—the generation of linguistic structures.[32] According to Marr, visual information is processed through a sequence of stages that begins with the focusing of incoming

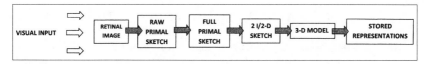

Figure 2.3 A theory of human vision developed by David Marr in 1982 stresses information processing and the recursive construction of mental visual representations, which seems to have been the foundation of the mind. (Modified from Michael C. Corballis, *The Lopsided Ape: Evolution of the Generative Mind* [New York: Oxford University Press, 1991], 220, fig. 9.1, based on David Marr, *Vision: A Computational Investigation into the Human Representation and Processing of Visual Information* [San Francisco: Freeman, 1982])

light on the retina, the light-sensitive cells at the back of the eyeball (figure 2.3). The retinal image is received in the form of a *raw primal sketch*, comprising a map of variations in light intensity; these variations are hierarchically grouped into larger units to form a *full primal sketch*. The next stage entails processing information about the shape, orientation, and distance of objects; Marr referred to it as the *2½-D sketch*.[33] In the final stage, or *3-D model representation*, the shapes and spatial arrangement of objects are organized hierarchically in an "object-centered coordinate frame."[34]

Molecular differences between living apes and Old World monkeys suggest divergence around 30 million years ago, although the earliest known ape fossils are dated to the beginning of the Miocene epoch (22 million years ago).[35] Bones and teeth assigned to the genus *Proconsul* from Africa dominate the fossil record of early apes. Their analysis reveals that *Proconsul* was a mixture of ape and monkey characters—an arboreal quadruped as large as a modern female gorilla and lacking a tail. The skull is relatively ape-like in appearance, with fully forward-facing orbits and a shortened snout. Brain volume is calculated at just below 170 cubic centimeters, which is high for apes in relation to the small body size.[36]

During the middle Miocene (between roughly 18 and 12 million years ago), the apes underwent a dramatic expansion of their range, spreading out of Africa and across western and southern Eurasia. This was facilitated by tectonics as the African plate collided with the Eurasian plate, creating a new land connection between the continents. The apes diversified into many new genera. Among the European taxa was *Dryopithecus*—first discovered in France in the 1850s—which exhibits more features in common with the living African apes than does *Proconsul*.[37] It is a plausible ancestor for the living African apes and humans and presumably possessed a brain and visual system fundamentally similar to that of both.

The Somesthetic Animal: Bipedalism and the Human Hand

[I]n the human race, a perfect tactual apparatus subserves the highest processes of the intellect . . . the most far-reaching cognitions, and inferences the most remote from perception have their roots in the definitely-combined impression which the human hands can receive.

HERBERT SPENCER

When Carolus Linnaeus published his systematic classification of plants and animals in 1753, he placed humans and the great apes in the same genus. Humans were classified as *Homo sapiens*, and chimpanzees and orangutans (gorillas were unknown to him) as *Homo troglodytes*. The decision to assign apes and humans to the same genus was based entirely on comparative anatomy and the already well-known morphological similarities between apes and humans. It did not reflect evolutionary relationships and was not a shocking idea.[38]

In the latter half of the nineteenth century, Darwin, Huxley, and others made the case for an evolutionary relationship—a common lineage—between apes and humans. This *was* a shocking idea for many people. In any case, the apes had been moved to another genus or genera and even into a separate family. In the twentieth century, as more of the human fossil record was recovered, the evolutionary divergence between apes and humans was eventually placed in the middle of the Miocene epoch, roughly 15 million years ago.

By the mid-1960s, however, studies of the molecular differences between living apes and humans suggested a much later split between the two. The fossils were subsequently reinterpreted and fell into line with the results of the molecular research. In more recent years, new molecular studies have yielded further evidence of the close relationship between humans and the African apes, placing both in a group apart from the Asian apes, or orangutans. (Humans have lost their classificatory status as a separate family.) And within the evolutionary clade that contains African apes and humans, chimpanzees and humans compose an even tighter group. This classification is based on both anatomical and genetic comparisons.[39] We have almost returned to where Linnaeus was in 1753.

The close relationship between chimpanzees and humans also is manifest in some aspects of behavior and ecology. In contrast to gorillas, chimpanzees forage over relatively open landscapes, engage in some hunting, and make a few simple tools. Their foraging adaptations and toolmaking, when combined with their anatomical and genetic overlap with humans, suggest that they probably resemble in many ways the last common ancestor of apes and humans. Despite this, both chimpanzees and gorillas evolved some features that probably were not present in this (still unknown) common ancestor. One of them is the unique mode of quadrupedal locomotion, *knuckle walking*, practiced by the African apes.[40]

There are inevitable gaps in the fossil record. Darwin attributed them to sampling and preservation bias—that is, "the imperfection of the geological record."[41] The problem may be further exacerbated by intervals of very rapid evolutionary change, or "punctuated equilibria."[42] There is nevertheless at least a broad evolutionary continuum from insectivores through prosimians and monkeys to apes and humans with respect to anatomy, genetics, and ecology. There are no massive gaps, for example, between Eocene prosimians and Miocene apes. But an immense gulf seems to lie between the cognitive world of the chimpanzee and the human mind—a difference that appears comparable to, if not greater than, that between prosimians and apes or, for that matter, between early chordates and placental mammals.[43]

How and why did the human mind emerge from the Miocene ape brain? How was the immense gulf crossed? Intermediate forms in the human fossil record fill much of the gap with respect to anatomy (and the genetic differences, as already noted, are not substantial). They include various fossil forms of *Homo* that reflect a continuum with respect to brain volume, which is undoubtedly a critical factor in the emergence of the mind. Equally important, there appear to be intermediate forms with respect to the mind, although they are more difficult to identify and interpret in the fossil record.

The short answer to the question is that early humans evolved a uniquely specialized extremity in the form of the hand, most likely an indirect consequence of *bipedalism*. The large neocortex and visual-motor system that had evolved over millions of years of primate evolution provided an essential context for this development. Humans eventually began to produce artifacts more sophisticated than those made by chimpanzees from which emerged the first externalized mental representations. They were relatively simple and exhibited little creativity, but they seem to have initiated a process—driven, I would suggest, by feedback between internal and external thought—that resulted in larger brains and more complex and creative representations. The social setting, the exchange of representations among members of the group, also must have been a factor in the process. The modern mind emerged when the ability to generate complex representations was expanded from hand to vocal tract in the form of syntactic language, creating the super-brain and unlimited creative potential.

Bipedal locomotion, or walking upright on two limbs, was the starting point, although for most of the time span that humans have walked the Earth, it had no significant consequences for brain volume. Bipedalism is equated with the evolutionary split between humans and African apes; it made the former out of the latter, and early humans have sometimes been described as bipedal apes. The early fossil record of humankind is very sparse, however, and the possibility cannot be excluded that the human lineage diverged from a chimpanzee-like ancestor for another reason and that bipedal locomotion came later. And it is also possible that bipedalism evolved among one or more extinct African apes that were not ancestral to humans. It is nevertheless the key feature that paleoanthropologists seek to identify among the few fossil remains of suspected human affiliation that are more than 3 to 4 million years old.

Bipedalism is an extremely rare form of locomotion among mammals, although like other features peculiar to humans it has roots in primate evolution—the upright posture found among apes, monkeys, and some prosimians. Most mammals are quadrupeds, and there are disadvantages attached to walking on the hind limbs. Around the world, humans struggle with the consequences in the form of knee-joint problems, back injuries, and related health issues. The advantages of bipedalism appear to be limited. A frequently cited study by Peter Rodman and Henry McHenry found that while it is less energy efficient than quadrupedalism at high speeds, bipedal locomotion is more efficient at slower (that is, walking) speeds.[44] Other possible advantages include reduced body-surface area exposed to the sun, improved ability to see predators in tall-grass habitats, and better capacity to carry objects.[45]

How and why did human bipedalism evolve? Although most paleoanthropologists believe that it had profound consequences for humankind—freeing the hands for improved toolmaking and all that followed—they suspect that it evolved for more prosaic reasons related to the immediate foraging challenges faced by Miocene apes. A widely held view, buttressed by Rodman and McHenry's study, is that bipedalism provided an energy-efficient means of foraging across landscapes with a low or medium density of trees. Climates became cooler and drier toward the end of the Miocene (roughly 8 million years ago), and many of the ape genera went extinct. In sub-Saharan Africa, forests shrank while grassland and open woodland expanded. The earliest humans may have evolved from *Dryopithecus* or a closely related African ape that spent an increasing amount of time foraging on the ground.[46]

The shift to bipedal locomotion affected many parts of the skeleton, not simply the limbs. In comparison with that in the African apes, the foramen magnum (through which the brainstem extends at the base of the skull to connect with the spinal cord) rotated forward, as the skull came to sit on top of the vertebral column, which, in turn, acquired an S-shaped curve. The pelvis became broader and shorter. The angle of the femur (thigh bone) was significantly altered. The lower limbs were elongated and the joint-surface areas expanded. The foot acquired a platform shape to support the upright body (figure 2.4). The changes in the bones were accompanied by changes in the musculature.[47]

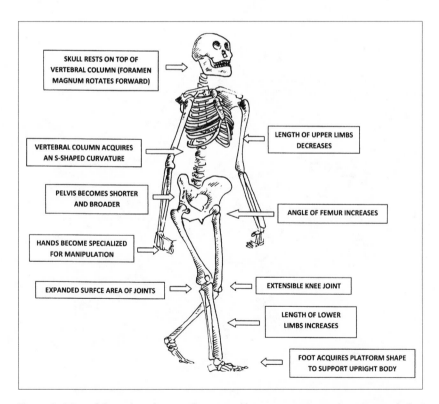

Figure 2.4 Bipedalism altered many elements of human anatomy and consequently has high visibility in the fossil record. The skeleton represents a reconstructed *Australopithecus afarensis*. (Redrawn from Richard G. Klein, *The Human Career: Human Biological and Cultural Origins*, 3rd ed. [Chicago: University of Chicago Press, 2009], 69, fig. 3.3)

The many changes in the skeleton mandated by the shift from quadrupedal to bipedal locomotion underscore the complexity of the transition,
which probably took place over an extended period and, initially, perhaps
was related to feeding posture.[48] The myriad anatomical changes generated
by the shift to bipedalism also yields a dividend for paleoanthropology—
high visibility in a sparse and fragmentary fossil record. There are many
potential signs of bipedalism in bones of suspected human affiliation dating to more than 3 to 4 million years ago, and paleoanthropologists search
among them for these indicators.

The oldest potential signs of bipedalism presently lie in fossils dating
to roughly 6 to 7 million years ago from west-central Africa assigned to
the newly created taxon *Sahelanthropus tchadensis* (figure 2.5). The fossils

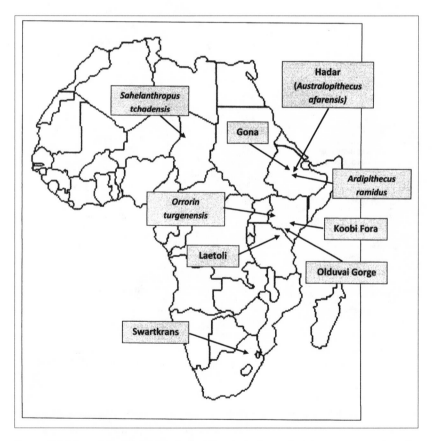

Figure 2.5 Map of fossil localities and archaeological sites in Africa, mentioned in the
text, that date to between 7 and 2 million years ago.

include a nearly complete cranium, lower jaw fragments, and teeth. The shape of the cranium had been distorted by geologic processes, but was restored visually to its original form by means of software technology. The short face, small canine teeth, and thick enamel on the teeth are human (as opposed to modern ape) characteristics. Most important, the position of the foramen magnum may reflect bipedal locomotion.[49]

The next oldest possibility dates to about 6 million years ago from northern Kenya. Fossils assigned to another new taxon, *Orrorin tugenensis*, comprise several limb bone fragments and teeth. In this case, the focus has been on the morphology of the femur, which exhibits a human-like angle an and elongated neck.[50] Computerized tomography scans may indicate a human-like distribution of cortical bone thickness, but this is controversial.[51] More recent fossils from Ethiopia, dating to between 5.7 and 4.4 million years ago, are classified as *Ardipithecus*. The youngest of these, assigned to *Ardipithecus ramidus*, include postcranial remains that reflect a primitive form of bipedalism.[52]

More human-like locomotion is found in the earliest fossils assigned to *Australopithecus*. Among the postcranial bones from two sites in Kenya dating to 4.2 to 3.9 million years ago and assigned to *Australopithecus anamensis* are a humerus (upper arm bone) and a tibia (lower leg bone) of a bipedal human.[53] Especially powerful evidence of upright walking comes from Laetoli (Tanzania), where footprints—presumably made by australopithecines—were preserved in volcanic ash deposits dating to about 3.6 million years ago.[54]

Between 4 and 2 million years ago, australopithecines dominate the human fossil record. These early humans diversified into several species that occupied various ecological niches in sub-Saharan Africa, and one of them apparently was the direct ancestor of the genus *Homo* and modern humans. The early australopithecines are best known from the "Lucy" skeleton, discovered at Hadar (Ethiopia) by Donald Johanson in 1974 and eventually classified as *Australopithecus afarensis*. Lucy is roughly 3.2 million years old and represents a small ape-like human that walked upright but retained some striking arboreal adaptations: long chimpanzee-like forearms, a grasping foot, and curved fingers. Lucy evidently spent much of her time in trees. Her brain remained small (400 to 500 cubic centimeters) and only slightly larger, relative to her body size, than that of a modern chimpanzee.[55]

The most important developments in human evolution during this period concern changes in the anatomy and function of the hand. Until the appearance of the earliest *Homo* roughly 2.5 million years ago, there were no significant increases in brain size above that of the African apes, and there are no known stone artifacts. But changes in the morphology of the hand reflect a protracted transformation from a Miocene ape to a modern human hand, which was apparently completed by about 1.5 million years ago. This development—as critical as any to the emergence of the mind—was a logical consequence of bipedalism, which freed up the forelimbs for novel functions.

The evolved human hand is a remarkably sensitive, versatile, and precise instrument for translating internal thought to the external world.[56] As Vernon Mountcastle observed in *The Sensory Hand*: "Movements and positions of the hand . . . are produced by the actions of 35 muscles (or 44 if one takes into account the three muscles that each have four insertion tendons) that control the movements of the 27 bones of the hand at their joints."[57] The movements of the hand are coordinated in the neocortex with the equally remarkable system of color vision that evolved among the catarrhines. And the hand functions not only as an instrument for effecting unusually sophisticated and subtle motor responses directed by the brain, but also as a sensory organ that gathers a special category of information from the external world. In this capacity, it is supported by a dense complex of nerve fibers in the glabrous (hairless) skin of the palm and fingers.[58] Among humans, the hand acquired sensory functions that seem to parallel those of catarrhine vision, providing three-dimensional hierarchically structured representations of the external world through the tactile sense. Humans have become a *somesthetic*, as well as a visual, animal.[59]

The human hand is anatomically unique. As a consequence of bipedalism, the human hand is fully specialized for manipulation. The anatomist John Russell Napier (1917–1987) noted that the hand plays a locomotor role only in early childhood, when it is used like a monkey "foot-hand" for quadrupedal movement. Napier emphasized the primary importance of finger–thumb *opposition*, which he considered "probably the single most crucial adaptation in our evolutionary history": "Opposition is a movement by which the pulp surface of the thumb is placed squarely in contact—or diametrically opposite to—the terminal pads of one or all of the remaining digits."[60] It is made possible by rotation of the thumb and facilitated by a thumb that is long in relation to the other digits. Among primates, humans have evolved the highest ratio of thumb length to index finger length.[61]

Napier subdivided human hand movements into nonprehensile and prehensile. Nonprehensile movements include "pushing, lifting, tapping, and punching movements of the fingers, such as typewriting or working the stops of a musical instrument."[62] Prehensile movements, such as gripping or pinching action between the fingers and thumb, include two primary and two subsidiary grips. The primary grips are the *precision grip* between the terminal pad of the thumb and the tips of one or more fingers, and the *power grip* between the fingers and the palm with the thumb providing reinforcement. The subsidiary grips are the *hook grip*, with the terminal joints bent acutely, and the *scissor grip* for grasping objects between the fingers (figure 2.6). The specialized skin of the hand, with its papillary ridges and sweat glands, also plays an important role in the grasping and manipulation of objects.[63]

While the range of hand movements is expanded by movements of the wrist, forearm, elbow joint, and shoulder joint, and even other parts of the skeleton, the basic movements, as Napier noted, are few.[64] But the variety of combinations of movements and the variety of material structures that

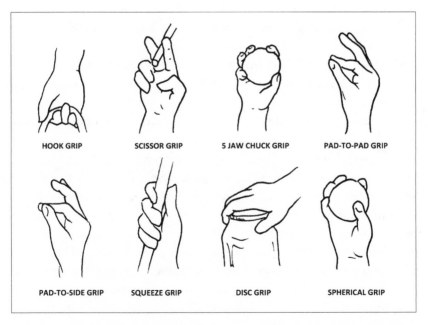

HOOK GRIP   SCISSOR GRIP   5 JAW CHUCK GRIP   PAD-TO-PAD GRIP

PAD-TO-SIDE GRIP   SQUEEZE GRIP   DISC GRIP   SPHERICAL GRIP

Figure 2.6 The basic grips of the human hand as defined by the anatomist John Russell Napier. (Redrawn by Ian T. Hoffecker, from Leslie Aiello and Christopher Dean, *An Introduction to Human Evolutionary Anatomy* [London: Academic Press, 1990], 372, fig. 18.1)

may be generated by hand movements coordinated by the modern human brain are potentially unlimited. The range of basic movements is analogous to the range of sounds that humans can produce with the vocal tract; it is a set of building blocks for a potentially infinite array of hierarchically organized technological creations. And through gesture, drawing, and writing, the role of the hands has been expanded to one similar to that of the vocal tract—the externalization of symbolically coded representations.

Like the other unique elements of human anatomy, the hand has deep evolutionary roots in the primates. The living primates are among the few groups of mammals that carry food to their mouths (as opposed to bringing their mouths to food),[65] and as early as the Eocene, the fossil ancestors of modern prosimians had evolved grasping hands with nails instead of claws. Finger–thumb opposition appeared in the catarrhines, which eventually evolved a variety of specialized hand adaptations—all but one of which combines locomotor and manipulative functions.[66]

Fossil bones of the early Miocene African ape *Proconsul* reveal a hand that, like other aspects of its anatomy, combines typical ape and monkey features (figure 2.7).[67] By the later Miocene (roughly 9.5 million years ago), a generic ape pattern is evident in hand bones of *Dryopithecus* from western Eurasia.[68] The African apes subsequently evolved their own specialized hand, with shortened thumb and elongated fingers (for suspending from branches) and other features related to their peculiar knuckle-walking locomotion.[69] The later fossil record of African apes is extremely sparse, however, and the sequence of events that yielded the modern ape hand is unknown.

The early human fossil record also is sparse and fragmentary, and there are only isolated and rather ambiguous specimens (no hand bones) dating to the interval between 8 and 4.5 million years ago. The 4.4-million-year-old hand and wrist bones of *Ardipithecus ramidus*, only recently unveiled in print, exhibit some novel features related to arboreal locomotion, but little indication of later trends.[70] The earliest indication of subsequent changes appears 3.2 million years ago in East Africa with *Australopithecus afarensis*, represented by Lucy. Among the remains recovered from Hadar (Ethiopia) are more than fifty hand bones, including carpals, metacarpals, and phalanges (fingers).[71] Analysis of these bones revealed many similarities to those of the living apes, including slender apical tufts (fingertips) on the ends of the phalanges. It also revealed some features lacking in apes but found among later humans, most notably a relatively long thumb (50 percent of the length of the middle digit). Lucy probably was capable of

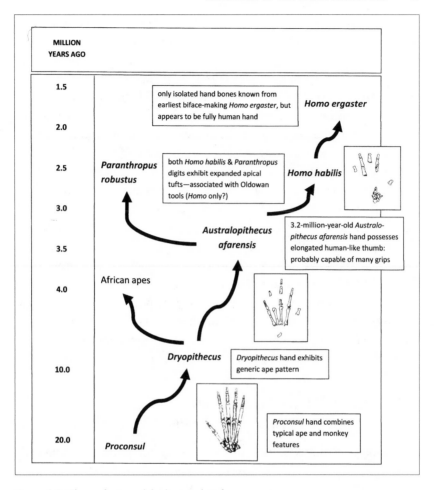

Figure 2.7 The evolution of the human hand.

some, but not all, of the grips and other hand movements performed by modern humans.[72]

The changes in hand morphology and function that took place 3.2 million years ago were not accompanied by an increase in brain volume or evidence of stone-tool production. But after 2.6 million years ago, there were increases in brain size (in both earliest *Homo* and the robust australopithecine *Paranthropus*), and stone tools are found in East Africa. Hand bones of early *Homo* associated with stone tools and dating to at least 1.75 million years ago were recovered in 1960 by Louis and Mary Leakey at Olduvai Gorge (Tanzania). The bones were described by Napier himself, who noted the expanded apical tufts (broad fingertips), similar to those of modern human fingers. But the Olduvai hand retains some ape-like features (for example, the morphology of the proximal and middle phalanges), and Napier wondered if it had been capable of a full precision grip.[73]

Of comparable age to the Olduvai finds are hand bones from Swartkrans (South Africa), also associated with stone tools and tentatively assigned to *Paranthropus robustus*. These hand bones differ from the earlier australopithecine pattern observed at Hadar and exhibit a number of human-like features, including expanded apical tufts like those of the Olduvai specimen.[74] It is not certain, however, that these hand bones belong to *Paranthropus*, since some remains of *Homo* also are present in the deposits.[75]

The fully modern human hand seems to have evolved by roughly 1.5 million years ago in *Homo* (both *Homo ergaster* and *Homo erectus*), although the supporting fossil evidence remains somewhat elusive. A few hand bones were recovered with a partial skeleton of *Homo ergaster* from Nariokotome (Kenya), which is dated to about 1.6 million years ago.[76] While these isolated specimens provide a less complete portrait of the hand than do the bones from Hadar and Olduvai, they are essentially modern in size and shape (like the arm bones also found at Nariokotome) and probably reflect the arrival of the fully modern human hand.

From Pebble Tools to Bifaces:
The Emergence of External Thought

Acheulean hand axes and other bifaces are a logical development from preceding Oldowan bifacial choppers.
RICHARD G. KLEIN

The turning point for humans took place between 2.6 and 1.6 million years ago in sub-Saharan Africa. It began with the earliest known stone tools (pebble tools and flakes) and ended with the emergence of externalized mental representations in the form of ovate bifacial artifacts, or hand axes. The first bifaces were simple and crude, barely distinguishable from some of the more heavily flaked cobbles and fragments of the preceding era. But within a few hundred thousand years, they became more refined; over the next million years, some hand axes were shaped with admirable precision (especially when the quality of the raw material permitted fine flaking control). The initial appearance of the bifaces is an event of fundamental importance for the origin of the human mind because it marks the beginning of a relationship between internal and external representations.[77]

The hand axes emerged, I would suggest, as a consequence of the inter-play between the brain and the stone tools that humans began to make 2.6 million years ago, mediated by the sensory hand and internal visual representation. Most archaeologists do not recognize a gradual transition from the pebble-tool industry (termed *Oldowan*) to a stone-tool industry containing hand axes.[78] The bifacial tools simply appear in the archaeologi-cal record 1.6 million years ago in Africa, marking the beginning of the *Acheulean* industry. But even a cursory glance at Oldowan artifact assem-blages reveals that the technique of bifacial flaking—shaping a piece of stone by chipping it on both sides—can be found throughout the history of this industry, including the earliest known assemblages.[79] Many of the partially bifacial artifacts of the Oldowan (some of which are classified as "proto-bifaces") resemble crude hand axes to some degree and simply lack a discernible ovate form.[80]

It is impossible to verify the thesis that hand axes were a consequence of interaction between the early human brain and Oldowan toolmaking, but it seems to be the most logical and parsimonious explanation of their origin. An alternative explanation would be that the production of bifaces was triggered by an event outside the realm of the Oldowan brain and its tools—for example, a previously untapped food source that required an ovate bifacial tool for successful exploitation. Even in this case, however, the earlier history of Oldowan toolmaking probably would constitute an essential prerequisite to hand-ax manufacture. Moreover, I am not con-vinced that the ovate design of hand axes was critical to their function. The analysis of microscopic wear patterns on their edges indicates that many bifaces were used as tools, but it has yet to be demonstrated that the same tasks could not be performed with Oldowan tools.[81]

The appearance of the bifaces 1.6 million years ago also marks the emer-gence of what I have called the proto-mind. The term is applied to the cognitive apparatus of the humans who made bifacial artifacts before the appearance of anatomically modern humans and evidence of modernity or the modern mind (which occurred after 250,000 years ago). Most, if not all, of these humans can be assigned to *Homo ergaster* and a younger, larger-brained species often classified as *Homo heidelbergensis*. They evolved the unique ability to externalize mental representations in the form of stone artifacts—as well as probably other artifacts made from materials that have not been preserved in the archaeological record—but lacked the cre-ativity of modern humans. They continued to produce the same external representation, the ovate biface, with only minor variations for more than a million years. In another respect, however, the proto-mind was similar to the modern mind: it was a collective phenomenon that entailed the sharing of one or more mental representations among individuals. The proto-mind was almost certainly a requisite precursor and stepping-stone to the modern mind.

If the identity of the makers of Acheulean bifaces is well established, the identity of those who made the Oldowan artifacts remains controversial. There is consensus that people assigned to the genus *Homo*—which first appeared at this time—made stone tools, but several other human taxa in sub-Saharan Africa are associated with Oldowan artifacts or evidence for artifact use. The remains of a gracile australopithecine (*Australopithecus garhi*) were found with mammal bones that exhibit traces of stone-tool damage, and it has been suggested that this species made the 2.6-million-year-old tools at nearby Gona (Ethiopia).[82] The remains of a robust australopithecine (*Paranthropus*) are associated with artifacts in younger deposits at Olduvai Gorge (Tanzania) and Swartkrans (South Africa), and the bones from Swartkrans included some from the hand.[83]

The importance of the human hand is underscored by some experimental research suggesting that much of Oldowan tool production lay beyond the abilities of the African apes. Beginning in 1990, several pygmy chimpanzees (bonobos) were trained to flake stone tools at the Language Research Center at Georgia State University. The apes struck cobbles with hammerstones and produced sharp flakes. Although many of their artifacts resemble Oldowan forms, comparison with the 2.6-million-year-old Gona tools revealed significant differences. The chimpanzees were unable to produce flakes of the same quality or quantity from a cobble.[84] Their brains are of comparable size to those of the australopithecines and not much smaller than those of earliest *Homo*, so the more constrained movement of their hands probably accounts for some of the contrast. Another factor related to bipedalism is that, from the outset, the quality of the Oldowan tools reflects the selection and transport (that is, carrying in the hands) of high-quality stone from its sources.[85]

The early Oldowan sites are found in eastern Africa, and, after 2 million years ago, they also appear in southern Africa. By 1.8 million years ago, Oldowan sites are present in southwestern Eurasia; sites containing similar artifacts are known from northern China by about 1.7 million years ago.[86] In eastern Africa, Oldowan sites are typically found along the banks of streams and shores of lakes. In southern Africa, they have been discovered in limestone cave fillings that may contain a mixture of human occupation remains, carnivore accumulations, and debris simply washed in from the surrounding area. At the time, African landscapes were becoming drier, with more extensive grassland and fewer trees, while animal life was undergoing significant change (that is, extinctions and new species).[87]

The interpretation of the Oldowan artifacts has been debated for years. The most widely used approach to classify the tools was proposed by Mary Leakey. Reflecting a traditional functional typological interpretation (at least with respect to many of the artifacts), she assigned the artifacts to categories such as *chopper* and *scraper*.[88] In 1985, Nicholas Toth published a classic paper on Oldowan artifacts that emphasized the processes of manufacture over the definition of formal types. Toth suggested that formal "types" may not have existed in the minds of the toolmakers and that the Oldowan typological classification had been imposed on a continuum of fractured cobbles and flakes (figure 2.8).[89] These seemingly arcane discussions of artifact classification among archaeologists carry heavy implications for human cognition. If the Oldowan tools were not shaped in accordance with mental representations (or "mental templates"), the bifaces that emerged after 1.6 million years ago—which almost certainly *do* reflect such a template—constitute a fundamentally new and different phenomenon.

The Oldowan industry comprises (1) *percussors*, or heavy stones used to strike other pieces of stone; (2) variously shaped *cores*, from which flakes were struck; and (3) the *flakes* themselves, which also vary in size and shape. Many artifact assemblages also contain (4) *retouched flakes*, or flakes that were chipped along the edges. According to the traditional classification, retouched flakes are often assigned to functional type categories such as *awl* and *side-scraper*, but also may simply represent variations on a continuum of chipping. Nevertheless, retouched flakes become more common in Oldowan sites after 2 million years ago. Much of the spatial and temporal variation in the assemblages appears to reflect differences in the stone used for artifact manufacture or, more specifically, differences in how various stones fracture. Commonly used stone included volcanic lavas and ignimbrites, as well as quartz and quartzite; cherts and flints are less common.[90]

The making and using of Oldowan tools was driven by human ecology. The tools seem to have played a critical role in allowing early humans to occupy an ecological niche that entailed a significant consumption of meat, especially meat obtained from large mammals. Some sites contain

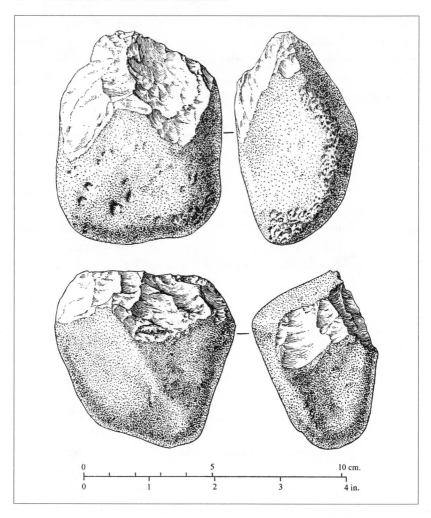

Figure 2.8 Stone tools of the Oldowan industry, classified as unifacial choppers, from Upper Bed I at Olduvai Gorge that do not appear to reflect the imposition of a mental template on a piece of rock. (From Mary D. Leakey, *Olduvai Gorge*, vol. 3, *Excavations in Beds I and II, 1960–1963* [Cambridge: Cambridge University Press, 1971], 77, fig. 40. Reprinted with the permission of Cambridge University Press)

the bones of various large mammals, many of which exhibit traces of damage from cobbles or heavy tools (used to smash them open for marrow) and/or sharp flakes (used to strip off meat).[91] Analysis of microscopic wear on the edges of tools from Koobi Fora (Kenya) revealed that some were used for cutting meat (although others yielded traces of woodworking and cutting soft plant materials).[92] And at least some of the large-mammal bones in Oldowan sites were derived from fresh carcasses, not leftovers abandoned by carnivores, and may reflect hunting or aggressive scavenging.[93] In sum, the Oldowan tools were made and used for very practical purposes, and they probably paved the way for early human expansion into new habitats and climate zones in Eurasia.

Many Oldowan sites have been discovered and excavated since 1971, when Mary Leakey published her volume on the artifacts from Olduvai Gorge. Nevertheless, the sequence of occupation levels at this most famous of early human sites still offers the best window on the emergence of external thought in the form of Lower Paleolithic bifacial stone artifacts. The occupation levels contain thousands of stone artifacts manufactured during the later Oldowan industry (roughly 2 to 1.5 million years ago) and the early Acheulean (roughly 1.5 to 1.2 million years ago). They represent the debris of human activities—utilized rocks, stone tools, and the waste products of toolmaking—scattered along the shore of an ancient lake that eventually retreated under drying conditions.[94] As at many other Oldowan sites, the bones and teeth of mammals as well as other vertebrates are found with the stone artifacts. At least some of the large-mammal bones exhibit traces of tool damage and represent human food debris.[95] The remains of both early *Homo* (*H. habilis*) and a robust australopithecine (*Paranthropus boisei*) are found in the occupations.

The major units in the lower part of the sequence at Olduvai Gorge (2 to 1.2 million years ago) are, from the top down:

Upper Bed II		Acheulean (bifaces)
Middle Bed II (upper part)		Acheulean (bifaces)
Middle Bed II (lower part)		Oldowan
Lower Bed II	1.8 mya	Oldowan
Upper Bed I		Oldowan
Middle Bed I		Oldowan
Lower Bed I	2 mya	Oldowan

The Oldowan assemblages represent the younger phases of the industry and yield a healthy quantity of retouched flakes. The upper levels of Bed II contain assemblages—in addition to those assigned to the Acheulean—classified by Mary Leakey as "Developed Oldowan B" that include a small number of bifaces.[96] Although the oldest bifaces in Africa are currently dated to about 1.7 to 1.6 million years ago,[97] the earliest Acheulean levels in Bed II at Olduvai Gorge do not seem to be significantly younger and represent an early phase of the industry. The most common raw materials used in both industries are quartz and quartzite.

One reason that archaeologists do not perceive a gradual transition from Oldowan to Acheulean is that the traditional typological classification forces individual artifacts into discrete categories. But as Nicholas Toth observed, many Oldowan tools actually occupy places along a continuum of variation.[98] And while no formal bifaces have been identified among the assemblages below the upper part of Middle Bed II, there are many examples of bifacially flaked artifacts. They do not seem to reflect a mental template (although we can never be certain that this is the case), but they were partially and sometimes completely flaked; that is, there are no traces of cortex or the original weathered surface of the cobbles. In Lower Bed I, all the "choppers" were bifacially flaked to create a working edge, and some were completely flaked on all sides. Tools classified as *discoids* also exhibit bifacial working and sometimes complete flaking, creating a symmetrical appearance. And in what may represent a significant trend, *proto-bifaces* ("intermediate between a biface and a chopper") become more common in the upper levels (figure 2.9).[99]

The bifacial tools in the early Acheulean levels at Olduvai Gorge and elsewhere are relatively crude in appearance—some are incompletely flaked and retain cortex—and the differences between them and many Oldowan tools are not dramatic. Nevertheless, when viewed as a group, the ovate artifacts in these and later Acheulean levels reveal a consistent pattern imposed on large flakes or cobbles that departs from the generally random shape of Oldowan tools (figure 2.10). They are subdivided into three types found throughout the Acheulean: hand axes, cleavers, and picks. All three exhibit an ovate form in plan view and cross section.[100]

The sequence of assemblages at Olduvai Gorge illustrates that humans were producing a variety of flaked-stone artifact shapes for hundreds of thousands of years before the emergence of formal bifaces. The experiments with the pygmy chimpanzees suggest that humans had evolved an unprecedented degree of control over the flaking process with their highly specialized sensory hands. Even more important, the analysis of stone artifacts from a 2.3-million-year-old site in Kenya indicates dynamic interaction between the brain of the toolmaker and the raw material. By refitting flakes and cores at Lokalalei 2C, Hélène Roche and her colleagues discovered that the Oldowan toolmakers were diagnosing—and correcting—

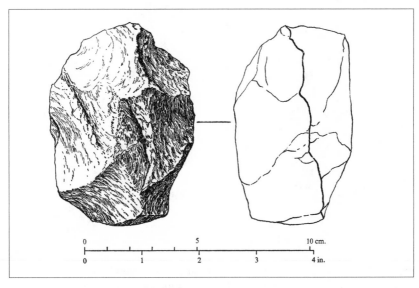

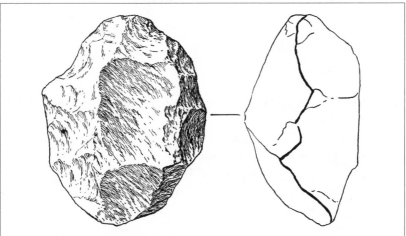

Figure 2.9 Stone tools of the Oldowan industry from the lower part of Middle Bed II at Olduvai Gorge that were flaked on more than one side. Classified as proto-bifaces, they seem to anticipate the flaking techniques used to produce the bifacial tools of the Acheulean industry. (From Mary D. Leakey, *Olduvai Gorge*, vol. 3, *Excavations in Beds I and II, 1960–1963* [Cambridge: Cambridge University Press, 1971], 101–102, figs. 51, 52. Reprinted with the permission of Cambridge University Press)

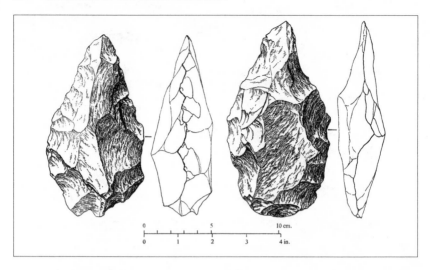

Figure 2.10 Bifacial tools, or hand axes, of the early Acheulean industry from Upper Bed II at Olduvai Gorge reflect an artificial or a semantic mental representation imposed on the rock. (From Mary D. Leakey, *Olduvai Gorge*, vol. 3, *Excavations in Beds I and II, 1960–1963* [Cambridge: Cambridge University Press, 1971], 206, fig. 97. Reprinted with the permission of Cambridge University Press)

imperfections in the surface of the area of the stone from which the flakes were struck.[101]

The ovate form, I suggest, was a result of the feedback between representations of the artifacts in the brain and the artifacts themselves. Each time an individual flaked a piece of stone, he or she created a series of new visual and tactile representations of the artifact—a nonverbal form of talking to oneself.[102] Some early humans eventually reversed the process and worked the stone in accordance with an internal representation. It was a simple three-dimensional geometric form, initially imposed rather clumsily on pieces of rock, shared among individuals in space and time that became a social mental phenomenon.[103]

Why did early *Homo* start to make bifacial artifacts or, more specifically, bifacial artifacts shaped in accordance with a preconceived form? The question is a long-standing one among archaeologists, and a definitive answer has yet to be given. Most assume that there is a functional explanation for bifaces: they were made to perform one or more functions that could not be performed with Oldowan tools, or at least not performed as efficiently with Oldowan tools. This assumption is supported by microscopic traces of use-wear on the edges of some bifaces. It nests comfortably within an evolutionary biology framework for human evolution: bifaces were adaptive and enhanced the long-term reproductive success of the humans who made and used them.

A variety of functions have been proposed. A widely held view is that bifaces were used primarily to butcher large mammals. The heavy tools seem to be particularly suitable for chopping through limb joints, and there is supporting evidence from both microwear studies and the repeated association of bifaces with the remains of butchered carcasses. Another suggestion is that they were used to dig through hard soil to obtain plant roots or other buried resources. Some archaeologists argue that bifaces were multipurpose tools—that is, Lower Paleolithic Swiss Army knives—or that they also served as cores, a portable source of fresh flakes.[104]

One proposed function that emphasizes the shape of bifaces is that they were used as projectiles. According to this view, their design reflects aerodynamic considerations and is tied to the evolution of throwing. It has been advocated with particular vigor by the neuroanatomist William Calvin, who argued that the neural pathways involved in throwing objects provided a basis for the subsequent evolution of syntactic language abilities.[105]

The initial appearance of bifaces in Africa was broadly coincident with some changes in human anatomy and ecology. The remarkable skeleton from Nariokotome (Kenya) dates to about 1.6 million years ago, and a lower jaw was found with early Acheulean artifacts at Konso (Ethiopia). These fossils belong to *Homo ergaster* (sometimes they are assigned to the closely related *Homo erectus*). They reflect an increase in brain size—almost 30 percent larger than the mean for *Homo habilis*—although the significance of this is mitigated by an almost equal increase in overall body size. More important, *Homo ergaster* was essentially modern from the neck down. The tree-climbing adaptations of the australopithecines and earliest *Homo* are absent, and the hand may be fully modern as well.[106]

The changes in human anatomy are almost certainly tied to changes in human ecology. *Homo* had expanded into new regions and climate zones in North Africa and mid-latitude Eurasia—as far as 40° North. Although climates were warmer than those of today, the movement north and out of Africa exposed humans to environments that were cooler, drier, more seasonal, and less biologically rich than those of tropical and subtropical Africa. Food was more dispersed in space and time. One response to this is an increase in foraging range; recent human foragers living in desert, tundra, or other resource-poor environments typically forage over very large areas during the year.[107] The changes in postcranial anatomy provided *Homo ergaster* with a more efficient walking adaptation that probably reflects the demands of wider foraging. Another response is greater reliance on meat, a more efficient food source than a vegetarian diet (recent human foragers in higher latitude environments rely heavily on meat). The analysis of large-mammal remains from upper Bed II at Olduvai Gorge suggests a more intensified focus on meat about 1.5 million years ago.[108]

The question is therefore twofold: Why did humans start to make formal bifaces, and how were these bifaces related to the changes in anatomy and ecology that took place at roughly the same time? On balance, the strongest case for biface function—supported by several lines of evidence—is that hand axes were used for large-mammal butchery. I think that it is important to bear in mind, however, that Oldowan tools were used effectively for the same purpose. And during the extended period when some humans were making bifaces in Africa and parts of Eurasia, other humans—especially in East Asia—continued to butcher large-mammal carcasses with pebble tools and flakes. Although clearly suitable for butchery, bifaces do not seem to have been essential for this task, and it is difficult to see how they conferred an advantage on those who made and used them. Moreover, the unworn edges of many Acheulean bifaces reveal that a high percentage of them were either not used or used only lightly.[109]

Bifacial artifacts seem to have played no role in the initial expansion of *Homo* into Eurasia. The earliest sites in this part of the world were occupied by people who made Oldowan tools. Formal bifaces did not appear in Eurasia until several hundred thousand years after humans had invaded new habitats outside the tropical zone. The oldest site outside Africa is Dmanisi (Republic of Georgia), where humans made Oldowan artifacts 1.8 million years ago. Their skeletal remains reveal that they had already evolved the modern bipedal anatomy of *Homo ergaster* while retaining the small brain and body of *Homo habilis*.[110]

In sum, I would conclude—until a stronger case can be made for their technical and ecological significance—that the function of bifaces as tools was a secondary one. And when they were used as tools, I suggest, their symmetrical design contributed little to their function. The bifaces are not what they seem to be, and evolutionary biology does not provide an adequate framework to explain them. They are based on artificial representations generated in the brain, apparently by the mental manipulation of visual and tactile representations received from the external world through the eye and sensory hand. They are the product of daydreams. As such, they are a manifestation of the "emergent properties" of the mind alluded to in chapter 1 and probably cannot be fully accounted for in terms of the principles of evolutionary biology. Only a *philosophy of mind* perspective allows us to appreciate their significance.

The appearance of formal bifaces in the archaeological record 1.6 million years ago strikes me as a major event in Earth history. It is analogous to the origin of life itself, which has no function in the context of the natural world; life is dynamic structure generated according to a set of properties that emerged with the formation of complex organic molecules in Precambrian times. The bifaces also reflect the emergence of new properties—in this case, arising out of the recently enlarged human brain, or what I have termed the proto-mind. If they have any analogy to the products of modern human culture, it might be to music, which also lacks any obvious function (although it has acquired various social and economic roles among modern human groups).

Making bifaces nevertheless requires both time and energy, and they must have conferred some advantage on their makers—even if subtle and indirect—in order to "pay for themselves" in the context of human evolutionary biology. Otherwise, it seems that the cost of making them, especially when so many were either unused or under-used, would have conferred a disadvantage on their makers. Not only was biface making a habit that lasted for more than a million years, but the makers and/or the habit spread geographically—from Africa to the Near East by 1.4 million years ago. It is necessary to explain why humans who found the visual-tactile structure of a biface so pleasing prospered and multiplied.

Looking ahead in time, it is clear that the externalization of mental representations, especially after humans evolved the capacity to create increasingly complex representations in the form of technology, eventually conferred immense advantages. By externalizing complex mental representations in the form of language and other symbolic media, members of a group would eventually share complex thoughts among themselves, creating a super-brain with unlimited generative powers, which had extraordinary consequences for them as living organisms. All subsequent developments in the emergence of the mind took place among the direct descendents of the biface makers. These developments provide some insight into why the making of bifaces more than a million years ago may not have been a waste of biological time.

While the bifaces themselves may have lacked practical value, daydreaming about bifaces may have been only one manifestation of the formation of artificial mental representations, some of which may have had very practical value. Biface making may have been intimately linked to an array of other daydreams externalized in the form of artifacts not preserved in the Acheulean archaeological record. They could have included tools and devices fashioned from plant and animal material that had some concrete benefits to their makers—for example, containers for carrying objects. Some of the microscopic traces of edge-wear from soft plant parts, wood, and hide found on Acheulean stone artifacts may reflect the making of such items.

More generally, thinking about and making bifaces may have been connected to the ability to think about something outside the immediate present in space and time—to create a visual representation, for example, of an object or a place that is not the product of processing incoming signals from the retina. The practical consequences of this ability, even in a limited proto-mind form, could have been profound, especially for humans foraging over wide areas for foods that were becoming increasingly variable in both space and time. It would have helped to imagine situations, predict future events, and plan ahead.

Making bifaces also may have had some impact on social relations within and perhaps even between groups because humans were not only externalizing a representation with their hands, but also communicating it to all other humans who could receive that thought by processing signals from the retina. And one of the most striking characteristics of Acheulean bifaces is their sameness; everyone expressed the same thought. This may have had some impact on social relations—acting as a binding agent within groups—which, in turn, may have had an effect on foraging success. Other mental representations with a potential cohesive influence on social relations may have been expressed not only in materials lacking visibility in the archaeological record, but in nonmaterial form altogether—through vocalization, gesture, or dance.

Landscape of the Proto-Mind

[I]t would be difficult to overemphasize just how strange the handaxe
is when compared to the products of modern culture.

THOMAS WYNN

Evolutionary biology is not the only interpretive framework applied to
Acheulean bifaces. Many archaeologists also have viewed them from a
culture-historical perspective, assuming that bifaces were made according to
a stylistic convention shared among people and communicated through
generations as a cultural tradition. The archaeological literature contains
many references to "Acheulean culture."[111] Unlike evolutionary biology, the
culture-historical perspective is predicated (at least implicitly) on a philosophy of mind. But it, too, suffers from shortcomings as an interpretive framework for the Lower Paleolithic because it is based on the modern mind.

The hand axes or bifaces are almost certainly the product of a mind
very different from our own. While unique among the animal kingdom,
the mind of the biface makers lacked some fundamental properties of the
modern mind. Much of the difference can be accounted for by the significant contrast in brain size. The cranial volume of the early biface makers
was between 800 and 1,100 cubic centimeters, which is roughly two-thirds
the size of the modern human brain. Their brains would have contained
billions of fewer neurons and an exponentially lower potential for synaptic
connections. The fossil crania of *Homo ergaster* nevertheless tell us little
about the pattern of thinking that lies behind the bifaces. It is the stuff
of the archaeological record that provides most of the insights into this
earlier form of mind that I term the proto-mind.

The archaeology required to explain the bifaces and the mind that created them is also very different from the discipline established in the nineteenth century to classify and interpret the artifacts of the modern mind.
It would be a variant of Colin Renfrew's cognitive archaeology, designed to
reconstruct or model a different mind on the basis of its external representations and their patterning in space and time. It is the sort of archaeology
that would be needed to understand traces of an extinct form of intelligent
life found on another planet.

Several major observations about the proto-mind emerge from the
archaeological record of the Acheulean. The first is obvious and already
discussed at length: the ability to create and externalize an artificial or a

semantic representation. As noted at the beginning of this chapter, it is unclear if the shape of bifaces was inspired by a natural object, such as a leaf (figure 2.11). It would seem that their design was a geometric abstraction, like a square, that is unknown in the external world—that is, outside the brain. Perhaps we need look no further than their immediate source: the bifacially flaked Oldowan tools. In any case, the ovate biface was created in the proto-mind and articulated by hand in analogical, or continuous, form; it was not constructed from discrete elements like phonemes or numerals (although the process of shaping it by removing individual flakes has a digital character).

The bifaces of the Acheulean reveal that the proto-mind was capable of thinking about something beyond its internal state. They demonstrate the property of what philosophers of the mind have (rather confusingly) termed *intentionality*, or "that capacity of the mind by which mental states refer to, or are about, or are objects and states of affairs in the world other than themselves."[112] Despite the possibility or probability that the structure of the biface was generated in the proto-mind with limited reference to the external world, the fact that this representation was projected outside the head is a simple but unambiguous illustration

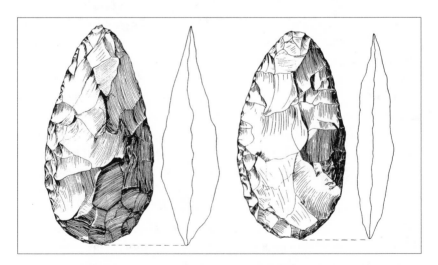

Figure 2.11 Bifacial tools of the Acheulean industry, dating to less than 500,000 years ago, from Kalambo Falls (Zambia) exhibit finer symmetry and shaping than their predecessors. (From J. Desmond Clark, *Kalambo Falls Prehistoric Site III: The Earlier Cultures: Middle and Earlier Stone Age* [Cambridge: Cambridge University Press, 2001], 337, fig. 6.6. Reprinted with the permission of Cambridge University Press)

of intentionality. In this way, the bifaces are analogous to a spoken or written sentence.

The second major observation is that the proto-mind was severely lacking in what Noam Chomsky identified as the "core property" of language—the ability to creatively recombine the elements of externalized mental representations to yield new structures in the same way that modern humans recombine words and sentences to create a potentially infinite array of meaning. The biface makers continued to produce essentially the same structure for more than a million years with relatively minor variation. It is as though they were speaking the same sentence over and over again with minor alterations in grammar and pronunciation.[113] There was no history because there was no historical process to create it, and there was no culture, at least in the sense that this term is applied to modern humans. The proto-mind was analogous to an asexually reproducing organism.

There was, nevertheless, *some* variation in the size and shape of Acheulean bifaces. In addition to the three most commonly recognized types (hand axes, cleavers, and picks), they include variants described as pointed, cordate, flat-butted, and ficron, among others.[114] The analysis of the differences may provide insight into not only the proto-mind, but also the emergence of the modern mind. Just as Nicholas Toth addressed the question of whether the Oldowan tools reflect the imposition of a mental template, or an internal representation, on flakes and cobbles of stone, some archaeologists have attempted to determine if the people who made the Acheulean bifaces had more than one "type" in their minds. Beginning in the 1960s, François Bordes, Derek Roe, Glynn Isaac, and other Paleolithic specialists applied various measurements and quantitative analyses to samples of bifaces.[115]

The results of these studies reveal a consistent pattern. Despite the common perception that Acheulean bifaces fall into several types, analyses of large samples have shown that the various forms grade into one another on continuous scales of size and shape. It seems that there was only one rather general representation—subject to what Isaac described as a "random walk" variation—in the proto-mind of the biface makers. Moreover, the variation seems to have been heavily influenced by the size and type of stone used and/or the degree to which flakes or cobbles were reduced by chipping. And it should be noted that the bifaces measured for these studies were derived primarily from later occurrences—archaeological sites that are less than a million years old—and that most probably were made by the larger-brained *Homo heidelbergensis*. Thus they do not reflect the early phases of the Acheulean.

In his analysis of bifaces at Olorgesailie (Kenya), Isaac noticed an interesting pattern. Although as a whole, the bifaces exhibit the same pattern of continuous variation, specific variants or modes tended to cluster together at particular locations (figure 2.12). At one locality, for example, most bifaces were small, narrow, and relatively thick; at another, they were of intermediate length, contained many chisel-ended forms, and exhibited minimal retouching. Isaac wondered if a single individual had produced most of the bifaces at any one locality or if members of one family or social group had coalesced around a particular design mode within the broader spectrum of variation.[116]

The same pattern has been noticed elsewhere[117] and probably tells us something about the workings of the proto-mind. It suggests a constant or recurring feedback relationship between the internal and external representations of bifaces. In addition to matching mental representations with the received visual and tactile representations of the bifaces they were making, individuals must have received the visual representations of bifaces made by others, especially in their immediate social setting. The feedback process reinforced individual or local trends that perhaps had begun as random variations, a sort of mental "founder's effect."

While the production of bifaces as a whole may have followed a random walk pattern, this apparently was not the case among individuals and/or local groups. Just as Oldowan people spent a million years tinkering with various forms of chipped stone and eventually matched internal and external representations of the ovate biface, Acheulean individuals and/or small groups toyed with the biface form (or "talked to themselves" in this way) for roughly the same length of time. There may have been other externalized representations not preserved in the archaeological record with which they developed a feedback relationship. I have suggested that the matching of internal and external representations by the Oldowan toolmakers underlies the emergence of the proto-mind; similarly, the feedback between internal and external biface representations may have contributed at least in part to the emergence of the modern mind.

The third major observation about the proto-mind is that, like the modern mind, it was a collective enterprise. The biface makers shared mental representations among individual brains, a phenomenon that had evolved in the honeybees but probably not among earlier vertebrates. In comparison with the iconic dancing of the honeybee, laden with vital economic information, the making of bifaces seems rather silly. But perhaps the chief importance of the bifaces was that they constituted shared structure—in

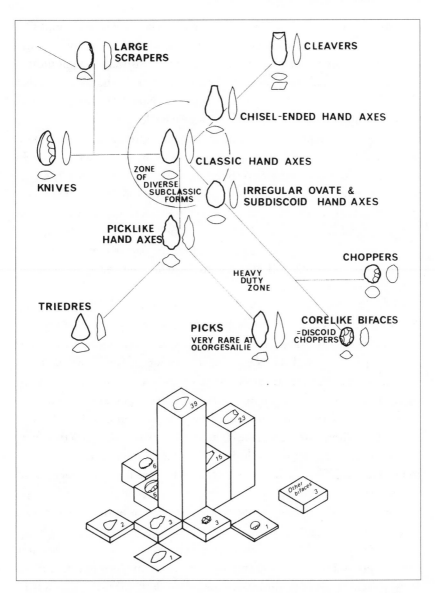

Figure 2.12 Variation in bifaces appears to be continuous and without clear division into subtypes, as observed by Glynn Isaac on the basis of his analysis of stone artifacts from Olorgesaile (Kenya). (From Glynn Ll. Isaac, *Olorgesailie: Archeological Studies of a Middle Pleistocene Lake Basin in Kenya* [Chicago: University of Chicago Press, 1977], 123, fig. 40. Reprinted with the permission of Barbara Isaac)

this case, "talking to one another"—regardless of content or meaning. This may have had a cohesive effect on social groups. In any case, the archaeological record shows that the biface representations were communicated through time as well as space, from one generation to the next, and that the collective Acheulean brain transcended individual organisms in a way that the shared representations of the honeybee do not.

A collective brain is a logical, if not inevitable, result of the evolution of the ability to externalize mental representations in a social setting. In addition to their effect on group cohesion, the bifaces also probably reflect a more dynamic cognitive environment. Even if the representations communicated back and forth within and between social groups were devoid of meaningful content, the simple fact of their existence—of the flow of mental representations among a group of individuals—seems significant. I would guess that it was critical to the subsequent emergence of the superbrain and modern mind.

Either the biface makers or the habit of biface making spread from Africa into the Near East roughly 1.4 million years ago. Acheulean bifaces occur at sites of this age in the Jordan Valley at 'Ubeidiya (Israel).[118] They did not spread beyond the Near East into other parts of Eurasia at this time, however. *Homo* populations in East Asia and the earliest known occupants of Europe did not make bifaces. As far as we know, there were no significant changes in human anatomy or ecology at this time.

After a million years ago, some changes are apparent in Africa that seem to have set the stage for the evolution of modern humans and the modern mind. The volume of the brain significantly increased, and a larger-brained form of *Homo* is recognized in the African fossil record by about 800,000 years ago. Often referred to as *Homo heidelbergensis*, reflecting the discovery in 1907 of a fossil jaw in Germany that appears to belong to the same species,[119] its estimated brain volume is between about 1,100 and 1,400 cubic centimeters and overlaps substantially with that of modern humans. *Homo heidelbergensis* retained primitive features of the skull, including a forwardly projecting face, thick supraorbital torus (brow ridge), and receding frontal bone (sloping forehead) (figure 2.13). Some paleoanthropologists nevertheless prefer to classify it as an archaic form of *Homo sapiens*.[120]

Coincident with and probably related to the changes in anatomy are changes in artifacts. Acheulean assemblages containing a high proportion of cleavers made on large flakes appear in East Africa after a million years ago (for example, at Olorgesailie [Kenya]), and similar assemblages in the

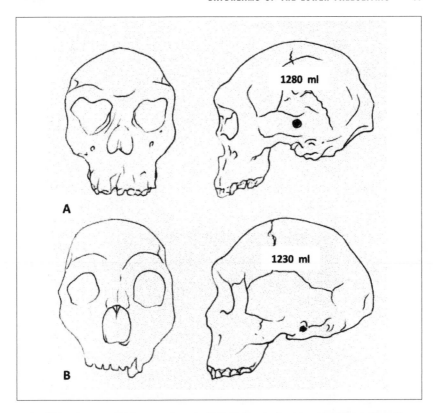

Figure 2.13 Crania often assigned to the later biface maker *Homo heidelbergensis* reflect a significant increase in brain volume (within the lower range of that of modern humans [mean = ~1,400 ml]): (A) Kabwe (Zambia) and (B) Petralona (Greece). (Redrawn from G. Philip Rightmire, "*Homo* in the Middle Pleistocene: Hypodigms, Variation, and Species Recognition," *Evolutionary Anthropology* 17 [2008]: 10, fig. 1)

Near East are dated to after 800,000 years ago (for example, at Gesher Benot Ya'aqov [Israel]).[121] Both human fossils assigned to *Homo heidelbergensis* and biface assemblages (often containing many cleavers) are found in Europe by 640,000 years ago and document the spread of the larger-brained *Homo* into higher latitudes (as far as 52° North in Britain).[122] The new biface makers also moved eastward into the Indian subcontinent, where they seem to be represented by both Acheulean artifacts and a partial cranium (Narmada) (figure 2.14).[123]

The workings of the proto-mind are very much in evidence during this great dispersal event and provide a sharp contrast to the subsequent migration, also out of Africa, of modern humans. There is evidence for

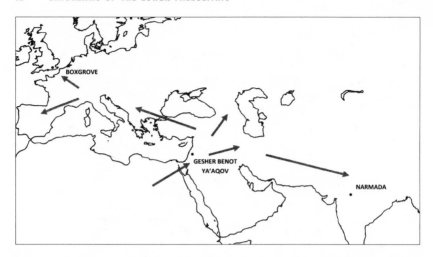

Figure 2.14 Although early biface makers occupied the Near East 1.4 million years ago, the major dispersal of biface-making humans (*Homo heidelbergensis*) out of Africa took place roughly 750,000 years ago, when they moved into the Near East, western and southern Europe, and southern Asia.

regional variation in the late Acheulean, and many bifaces are shaped with greater refinement and precision than those of the early Acheulean.[124] But the lack of creativity is manifest in an essentially uniform pattern of stone-artifact production, especially the bifaces. There is nothing comparable to the spatial and temporal differentiation of material culture that took place after 60,000 years ago when modern humans spread throughout the Old World.

3

MODERN HUMANS AND THE SUPER-BRAIN

For by Art is created that great Leviathan called a Common-wealth
. . . which is but an Artificiall Man.

THOMAS HOBBES

ARCHAEOLOGICAL EVIDENCE of the modern mind (or modernity) emerged in Africa after 100,000 years ago. It is associated with anatomically modern humans (*Homo sapiens*). At first, the signs of modernity seem rather modest; they are confined to perforated-shell ornaments, polished bone awls and points, and simple geometric designs incised into lumps of red ocher.[1] But roughly 60,000 years ago, some of these anatomically and behaviorally modern humans spread out of Africa and into other parts of the Old World, including Australia. In a relatively short period of time, the geographic area of the modern human archaeological record was expanded greatly, and at least some parts of this vast domain yield spectacular traces of modernity in the form of sculpture, painting, and musical instruments dating to more than 30,000 years ago.[2]

The archaeological record offers other items and patterns widely deemed to be evidence of modern behavior. The polished bone awls and points found in Africa before the dispersal of *Homo sapiens* into Eurasia are only the beginning of an impressive expansion in the variety and complexity of technologies. By 30,000 years ago, the presence of tailored sewn clothing, hand-operated rotary drills, high-temperature kilns, devices for harvesting small mammals, and other innovations are documented or reliably inferred from one or more regions occupied by modern humans.[3]

Other indications of modernity are said to include the production of stone blades (as opposed to flakes), the appearance of more standardized artifact types, the organization of domestic space, and the construction of complex hearths.[4]

A few years ago, the South African archaeologist Lyn Wadley, frustrated with the empirical "shopping list" approach to defining modernity in the archaeological record, suggested something different. Wadley argued that modernity should be defined with reference to a coherent theoretical framework. There is a widespread view that the use of symbols or "symbolically organized behavior" represents the fundamental difference between modern humans and their predecessors. Spoken language is the predominant mode of symbolic expression, but it seems to have no visibility in the archaeological record. Wadley therefore suggested that the "storage" of symbols in material form, such as visual art and stylistic attributes of artifacts, is the best way to identify modernity in archaeological terms.[5]

The use of external symbols is only part of what distinguishes the modern mind from the proto-mind, however. As noted in chapter 1, the linguist Noam Chomsky concluded that creativity (he used the more prosaic term "discrete infinity"), not the use of symbols, is the "core property" of language and that, following the evolutionary psychologist Michael Corballis, the same property can be extended to all forms of artificial or semantic mental representation. Creativity—the ability to recombine the elements of mental representations in hierarchical form, yielding potentially infinite variations of structure—is a unique and essential characteristic of the modern mind. It is the capacity to generate *alternative reality*.[6] While the use of external symbols is an important realm of creative expression, it is only part of a larger phenomenon. By excluding the nonsymbolic media through which the creativity of the mind is expressed, much of modernity is lost.

The most consequential nonsymbolic realm of creative expression is technology. After roughly 100,000 years ago, if not earlier, the technology of modern humans exhibits the same unlimited creativity recognized in language: a potentially infinite range of variation in hierarchical form with embedded components. From a relatively simple technological base comprising scrapers, bifaces, and other items of flaked stone; composite tools and weapons; and some shaped wooden artifacts, modern humans created an increasingly diverse and complex array of implements and devices. The growth of technology was cumulative. At 75,000 years ago, humans were shaping points and awls of bone. By 25,000 years ago, they had invented cold storage and fired ceramics. By 17,000 years ago, they were making simple mechanical devices, baking bread, and domesticating dogs.

Technology is not the only nonsymbolic avenue of creative expression among modern humans. Although it may have little impact on population and ecology (Steven Pinker described it as "useless" from an evolutionary biology perspective), music is another major realm of potentially infinite and often complex variation in structure.[7] Claude Lévi-Strauss, who labeled it "the supreme mystery" of humankind, observed that music possesses its own unique character that sets it apart from language and art as well as technology.[8] While symbolic, iconic, and especially stylistic elements may be incorporated into compositions, music is fundamentally not symbolic. In my view, it is an important part of the modern mind and should not be excluded from the archaeological definition of modernity. And it exhibits a surprisingly high visibility in the archaeological record after 40,000 years ago.[9]

There are other channels of creativity among modern humans, but their visibility in the archaeological record is either extremely low or ambiguous. Dance and cuisine are difficult to discern in the early prehistoric record, while the structuring of social roles—which acquires enormous significance in later times—is only tentatively identified in a Paleolithic context. These areas of creativity cannot be included in the archaeological definition of modernity at present, but they are part of the larger frame of reference and remind us that the offspring of the mind take many forms, not a few of which are nonsymbolic.

I suggest that the true significance of symbols, which appear in the archaeological record roughly 75,000 years ago, is that they probably indicate the formation of the super-brain (figure 3.1). The concept of the super-brain is analogous to that of the super-organism: multiple individuals perform functions normally confined to a single organism. In the case of a collective brain, multiple individuals gather and organize information about the environment. Information is shared by one individual with others, and some of it is stored and passed on from one generation to the next. Honeybee colonies represent a super-organism that contains a super-brain, and honeybees also gather, share, and process information about their environment among their members. The human super-brain embraces a vastly more complex and varied array of information, however, and it must have conferred significant advantages on its owners.

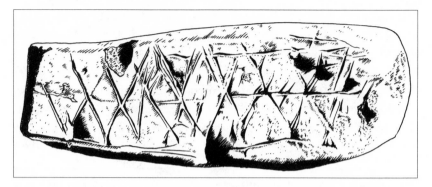

Figure 3.1 Simple geometric designs engraved on pieces of red ocher from Blombos Cave (South Africa), dating to roughly 75,000 years ago, are the oldest known examples of external thoughts in symbolic form (although perforated shell ornaments, which also may carry symbolic meaning, are known from even older contexts). (Drawn from a photograph by Ian T. Hoffecker)

Language is the primary means of integration of the human super-brain, although there are other means such as visual art. The representations of language are composed of symbols, and while the existence of symbols is as old as the metazoan brain, their presence *outside* the brain is extremely rare. A large-brained African *Homo* had evolved a means to externalize mental representations not in simple iconic form, but in true symbolic form (that is, information coded by arbitrary referents) as it exists in the brain; moreover, the structure of these representations was not constrained by materials such as wood or stone, like earlier forms of external thought. And although the archaeological evidence is confined to the rather clumsy and inefficient form of externalizing symbols with the hand, there is a widespread belief among paleoanthropologists that it is a proxy for spoken language—that the vocal tract had finally acquired a function analogous to that of the hand as a means to project mental representations outside the brain.[10]

In the early 1990s, Chomsky and his colleagues began to explore the idea that while language is poorly designed as a communication system, it may be optimally designed to interact with "the systems that are internal to the mind." The seeming imperfections of language (for example, the needless redundancy of noun and verb agreement) might be explained better from such a perspective[11] This notion (dubbed "The Minimalist Program") is compatible with the suggestion that language, as well as other forms of symbolic expression, represents thought moving from one brain to others in a manner analogous to information moving from one part of the neocortex to another. When humans share thoughts with one another, they are replicating the internal activity of the individual brain (which is a major portion of brain activity) on a larger scale.

Equally important, the archaeological evidence for creativity coincides at least broadly with the evidence for symbolism, which suggests that the super-brain was a prerequisite and a threshold for the generative mental powers of behaviorally modern humans. It is only with the appearance of symbols in the archaeological record that humans seem to take off—at first rather slowly—as a potentially unlimited creative force. The apparent linkage of the two produces a seeming dilemma, however. Spoken language, so widely inferred from the presence of symbols rendered by hand, is itself a primary manifestation of potentially unlimited creative ability. How could it also have been a prerequisite for the super-brain?

The super-brain may be the key to explaining other unique and fundamental aspects of human cognition, as discussed in chapter 1. These, in turn, seem likely to be related to language and symbolism. Higher-order consciousness and a "sense of self" are arguably tied to the use of symbols and the ability to create mental representations, in a variety of media, that contain oneself, as well as representations that contain oneself in other times and places. Such artificial representations are possible only with symbols. And from this, recognition of an "unconscious" self seems to follow. All these aspects of psychology, it seems to me, are plausibly linked to the phenomenon of the super-brain.

The *mind* is the super-brain: the integration of brains facilitated by language and other forms of symbolism created the mind. Although the super-brain is composed of a group of individual brains that are the product of evolutionary biology, it exhibits properties that are unknown or had not been evident in organic evolution. These properties apparently emerged as a consequence of the formation of a system of unprecedented complexity on Earth (and perhaps elsewhere in the universe). The core property is the generation of a potentially infinite variety of hierarchically organized structures of information in various media, analogous to the generation of potentially unlimited gene combinations through sexual reproduction. The super-brain / mind transcends biological space and time, accumulating and organizing information over the course of countless generations of organisms.

Emergence

> [P]arts of the brain (indeed, the major portion of its tissues) receive
> input only from other parts of the brain, and they give outputs to
> other parts without intervention from the outside world. The brain
> might be said to be in touch more with itself than with anything else.
>
> GERALD M. EDELMAN

The emergence of the mind is the central issue in paleoanthropology.
Archaeologists have framed the problem more narrowly, in terms of
behavioral modernity, defined in the archaeological record as evidence
for symbolism. Most of them seem to feel that language, represented in
proxy form by the symbolic content of artifacts and features, is the centerpiece of modernity. I have defined the issue more broadly, in terms of
not only the super-brain, which is identified with the same archaeological evidence for symbolism, but also creativity, which is documented with
a different sort of archaeological evidence in a wider temporal and spatial
context. However defined, it seems to me that the issue is fundamentally
the same.

If the origin of the mind is the central issue in paleoanthropology, it is
also the most challenging—a seemingly intractable problem. No one seems
to know how and why the mind developed, or (defined more narrowly)
how modernity and symbolism may be explained. Speculation about the
origin of language became so rampant after the publication of *On the Origin of Species* that the French Academy of Linguistics famously refused to
publish papers on the subject in 1866. A century and a half later, incredibly little progress has been made. Genetic mutations in speech function
and working memory have been proposed as a (random) trigger for modernity,[12] although we are still left wondering how and why the context for
such mutations could have evolved.

The eruption of Mount Toba on Sumatra roughly 75,000 years ago
may have been an environmental factor. The massive volcanic eruption
generated an ash plume that apparently encircled the globe, reducing temperatures by several degrees and devastating plant and animal life in many
regions. A few years ago, Stanley Ambrose suggested that the consequences
of the Mount Toba eruption may have forced the African *Homo* population through a severe evolutionary bottleneck, creating circumstances
under which rapid genetic change could take place as a result of drift,
founder's effect, or local adaptations.[13]

Yet another variable may have been competitive interactions between anatomically modern humans (*Homo sapiens*) and their Neanderthal cousins (*Homo neanderthalensis*). The dispersal of *Homo heidelbergensis* into western Eurasia after 1 million years ago eventually yielded a northern large-brained form of *Homo* that exhibited many of the same behavioral traits as its African contemporaries between 300,000 and 100,000 years ago. The Neanderthals seem to have expanded into the Near East at certain times (probably during intervals of cold climate), while modern humans were at least briefly present in the same region about 100,000 years ago. Thus the geographic distribution of the two species may have converged at times, leading to competition between two closely related and highly intelligent species of human with potential evolutionary consequences for each of them.[14]

Despite the absence of a compelling explanation for the origin of language and the mind, much is known about the developments in anatomy and behavior (some of which is manifest by preserved external thoughts) that preceded the appearance of symbolism and creativity in the archaeological record. These developments include a significant expansion of brain size, especially pronounced in certain areas of the brain (reflected in the size and shape of fossil crania). They may include changes in the vocal tract related to speech, but these may have occurred earlier. The changes in anatomical structure were accompanied by the ability to externalize representations of greater variety and complexity (that is, comprising more hierarchical levels). All these developments seem relevant to the subsequent arrival of symbolism and creativity.

Two general observations of a theoretical character are worth making at this point. One is that if the principles that govern creativity are emergent properties of the brain—a system of unprecedented complexity—this critical aspect of modernity may have been a consequence of developing complexity, rather than an adaptive trait that evolved because selection favored a more creative human organism. The second, and closely related, observation has been made by many students of human evolution. It is that at least some aspects of modernity or the modern mind may have evolved in another context, as features or systems designed for other functions, and only later reconfigured for functions associated with modernity. This phenomenon does not seem uncommon in evolutionary biology, and a frequently cited example is the reconfiguration of feathers among birds—thought to have evolved for thermoregulation—as part of their flight adaptation.[15]

A puzzling aspect of the paleoanthropological record is the seeming disconnect between the changes in anatomy and those in behavior that are associated with modernity. In terms of anatomy, modern humans are present in Africa by 200,000 years ago at the latest. But the archaeological evidence for symbolism so widely viewed as the sine qua non of modern behavior does not appear until about 80,000 to 75,000 years ago. In a recent essay, the paleoanthropologist Ian Tattersall observed that "in evolutionary terms this disconnect was entirely routine, for every new behavior has to be permitted by a structure that already exists."[16] And perhaps the gap is more apparent than real. There may have been critical anatomical changes in the brain that are simply not visible in the fossil skeletal record, while earlier archaeological evidence of symbolism and other aspects of modernity probably remain to be discovered.

The Information Animal

Instead of asking the standard functionalist question: is [language] well designed for use? we ask another question: is it well designed for interaction with the systems that are internal to the mind?
NOAM CHOMSKY

Anatomically modern humans evolved from a large-brained form of *Homo* in Africa roughly 250,000 years ago. This *Homo* is often assigned to the taxon *Homo heidelbergensis*. It expanded into Eurasia after 1 million years ago, establishing a separate northern lineage. The African lineage evolved, gradually it seems, into anatomically modern humans (*Homo sapiens*), who were present in sub-Saharan Africa no later than 200,000 years ago. The northern or Eurasian lineage evolved into the Neanderthals (*Homo neanderthalensis*) during the same time period. *Homo heidelbergensis* also may have had a shadowy presence in East Asia.[17]

The most important anatomical changes related to the emergence of modernity are alterations in the size and shape of the brain, and evolution of the modern vocal tract. The increase in brain volume is presumably important to the expansion of cognitive abilities (that is, the growing complexity and variety of the externalized mental representations observed in the archaeological record), while the change in the vocal tract is logically connected to spoken language and the use of symbols (that is, making artifacts with the larynx). Both carry evolutionary "costs" and require an explanation as to how their benefits outweighed their disadvantages in terms of natural selection. Large brains are expensive and place added demands on their bearers with respect to energy and nutritional needs.

And the configuration of the modern human vocal tract exposes its own-ers to a significant risk of choking.[18]

Contemporary theories of the mind seem to be based on the assump-tion that the size and complexity of the brain—the staggering number of neurons and synaptic connections—are essential to its function.[19] It makes sense, therefore, that archaeological evidence of fully modern behavior should be preceded by substantial growth in cranial volume. But volume among living and recent humans varies greatly, and there are many exam-ples of individuals with normal or above-normal intellectual abilities who have small cranial capacity (as low as 1,000 to 1,100 cubic centimeters).[20]

Between 500,000 and 200,000 years ago, cranial volume increased sig-nificantly among both the sub-Saharan and the north Eurasian humans, which complicates the question of why larger brains evolved. Why did the Neanderthals develop an oversized cranium similar to that of modern humans during the same period of time? Although both species exhibit signs of an expansion in the complexity and variety of external representa-tion, only modern humans seem to have made the final leap to external symbols, super-brain, and unlimited creativity. This suggests that large brain volume is a necessary but not a sufficient basis for the modern mind and that after a threshold size is achieved, a triggering event that may have little or no visibility in the fossil record also is necessary for the final leap (for example, a genetic mutation that occurred in modern humans but not in Neanderthals).[21] The parallel evolution of brain size also may have been driven by different factors in each species, although it seems an odd coincidence that cranial volume should reach roughly the same level at the same time in both species.

African skulls often assigned to Homo heidelbergensis date to between 600,000 and 300,000 years ago and include specimens from Bodo (Ethio-pia); Kabwe, formerly Broken Hill (Zambia); and Ndutu (Tanzania).[22] They possess a cranial volume of between approximately 1,100 and 1,300 cubic centimeters, which represents a significant increase over earlier African Homo, but falls short of the brain volume of modern humans (typically more than 1,350 cubic centimeters). By 200,000 to 150,000 years ago, anatomically modern crania are present in sub-Saharan Africa, including at Omo Kibish and Herto (both Ethiopia) (figure 3.2).[23]

In Europe, Homo heidelbergensis is perhaps represented best by the com-plete cranium from Petralona (Greece), which is very similar to the Kabwe skull but with a slightly lower volume of 1,230 cubic centimeters.[24] Even the earliest European specimens assigned to this taxon exhibit features that

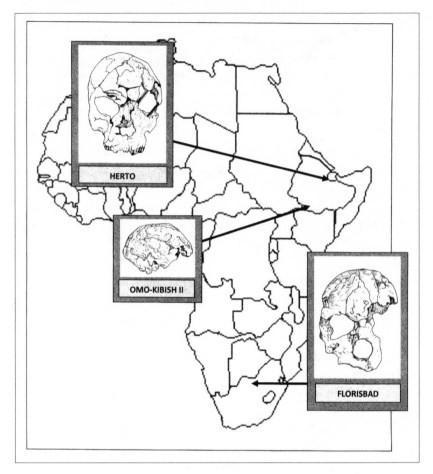

Figure 3.2 Anatomically modern humans had evolved in sub-Saharan Africa by roughly 200,000 years ago. (Map courtesy of the University of Texas Libraries, The University of Texas at Austin)

suggest a diverging northern lineage that eventually produced the highly distinctive Neanderthals. Mandibles from the Sima de los Huesos (Atapuerca) in northern Spain (probably more than 500,000 years old) exhibit a pronounced gap between the third molar and the ramus (*retromolar space*), which is a classic Neanderthal feature. The expansion of brain volume also is evident in this sample; one of the crania possesses an impressive volume of 1,390 cubic centimeters. The European Neanderthals later evolved a mean volume of slightly over 1,500 cubic centimeters, which was comparable to that of anatomically modern humans in Africa at the time and somewhat higher than the mean for living humans.

Cold climate may account for at least some of the increased size of the Neanderthal cranium. Many aspects of Neanderthal anatomy conform to a general pattern of cold adaptation among warm-blooded animals, including the comparatively short forelimbs and thick chest (reducing heat loss and protecting against frostbite).[25] Among modern humans, the largest cranial volume is found in populations that have occupied the circumpolar zone for several thousand years or more.[26] These populations also exhibit some of the other features found among the Neanderthals that are attributed to cold-climate adaptation. And, significantly with reference to brain size, these features often appear to have developed to an extreme degree in the Neanderthals (who have been described as "hyperpolar"), possibly reflecting their relative lack of technological protection against low temperatures.[27]

None of the increase in brain size in African *Homo sapiens* can be accounted for by climate, however, and an explanation must be sought in the increased cognitive demands of society, foraging, and/or technology. A hypothesis proposed by Robin Dunbar is that the modern human brain expanded in response to widening social networks. When these networks grew beyond the residential foraging group and became too large to maintain by grooming—an instrument of social bonding among higher primates—Dunbar suggests language evolved as a substitute for maintaining social ties ("vocal grooming").[28] An earlier and related idea (labeled the Machiavellian Intelligence hypothesis) is that large primate brains reflect the computational requirements of negotiating complex social systems.[29]

The "social brain hypothesis," which rests primarily on the observed quantitative relationship between the size of the neocortex and that of the social group among primates, seems plausible. But the archaeological evidence that accompanies the expansion of brain size (which includes both modern humans and Neanderthals) indicates significantly enhanced cognitive abilities related to the design of artifacts. The artifacts include more-complex hierarchically organized cores, composite tools and weapons with as many as four components, and more creative design variations in the shapes of stone implements. Occasional glimpses of less archaeologically visible materials suggest that these abilities were applied in other contexts; for example, preserved wooden spears from Germany reflect a surprisingly complex sequence of design steps.[30] I suggest that at least some of the growth in *Homo* brain size roughly 500,000 years ago is tied to the expanded multimodal processing, organizing, storing, and manipulating of information implied by these archaeological data.

And it would appear that the *Homo* brain was undergoing a general expansion of storage space that allows for one of the most striking attributes of living humans—the immense storehouse of memories and knowledge that each individual carries. This information certainly includes a large quantity of social data—details about the people with whom we regularly or periodically interact—but also comprises a vast amount of environmental data. One of the salient characteristics of recent hunter-gatherers is the large quantity of structured information about the landscape that each adult stores in his or her brain. It includes comprehensive and detailed classification systems of plants and animals, which is especially impressive in tropical settings, where the diversity of biota is very high.[31] The collection and storage of information is part of the environmental adaptation of all higher animals, but humans seem to have worked themselves into an ecological niche in which it became central.

More than any other organism that has ever trod the Earth, modern humans are the product of *information*, rather than genes (or, because genes are themselves a form of information, an alternative and more mutable type of information). To an unparalleled degree, an individual is the sum of an immense mass of integrated memories and knowledge accumulated over a lifetime of experience.[32] The quantity of information required to function as an adult is so great that roughly 25 percent of a human lifetime must be devoted to learning as a child and young adult, an excessive period of offspring dependence analogous to the grotesquely oversized cranium. All this information, much of it in the form of complex visual representations, requires literally billions of neurons organized into groups and networks. To accommodate it in such a large quantity, brain volume in African *Homo* expanded significantly after 1 million years ago (although the reason why it increased at this time is not clear), providing the basis for a novel form of organism, which—like the technology it was beginning to create—was largely constructed from nongenetic information.[33]

The African *Homo* brain also evolved a somewhat different shape during this period, and the altered proportions of various regions of the brain have been regarded as significant. Although the pronounced verticality of the modern human frontal bone suggests an expansion of the *frontal lobes*, comparative analysis of extant hominoids reveals that human frontal lobes remain proportionally the same as in the living apes. Within the frontal lobes, however, there is some percentage increase (6 percent) of the *prefrontal cortex*, which is almost certainly important.[34] Brain-imaging

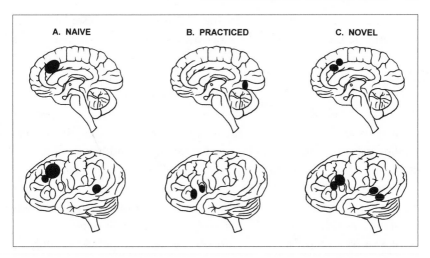

Figure 3.3 A critically important evolutionary change in modern human brain anatomy may have been the expansion of the prefrontal cortex, which appears—on the basis of recent neuro-imaging studies (PET scans)—to be the primary locus of generativity: (A) the prefrontal cortex is active when a subject is presented with a novel task; (B) activity declines when the task is a familiar one; (C) the prefrontal cortex becomes active again when a different task is introduced. (Redrawn from Elkhonon Goldberg, *The New Executive Brain: Frontal Lobes in a Complex World* [Oxford: Oxford University Press, 2009], 90, fig. 7.1, adapted from Marcus E. Raichle et al., "Practice-related Changes in Human Brain Functional Anatomy During Nonmotor Learning," *Cerebral Cortex* 4 [1994]: 8–26)

research points to the prefrontal cortex (interconnected with most other regions of the brain) as the center of creativity, or the ability to generate artificial representations (figure 3.3).[35] The most pronounced expansion in the modern human brain is seen in the *temporal lobe*, which is involved in language function.[36]

The evolution of the vocal tract, the other major anatomical change linked to modernity, presents a special challenge. The earliest humans presumably possessed a vocal tract similar to that of their immediate ape ancestors. Over the next few million years, it evolved into the vocal tract found in living humans. As in most mammals, the *larynx* (voice box) is positioned relatively high in the neck in the African apes, at the same level as the first three cervical vertebrae. In this location, it can shift position to separate the passageway to the lungs (windpipe, or *trachea*) from the tube to the stomach (*esophagus*). In living humans, the larynx migrates downward during infancy to a location opposite the fourth to seventh cervical vertebrae, where it cannot separate the windpipe from the esophagus (figure 3.4).[37] This creates the choking risk mentioned earlier, which accounts for several thousand deaths each year, but expands the size of the vocal chamber or *pharynx*. Using the vocal folds of the larynx, living humans can generate a wide range of sounds in this enlarged vocal chamber.

The morphology of the vocal tract also may be tied to other aspects of human evolution. The larynx functions to close off the windpipe during swallowing and lifting heavy objects; this retains air and helps solidify the trunk as a steady frame for the more demanding arm movements. There are some parallels between humans and gibbons (lesser apes) with respect to larynx function, and this may be significant not only because gibbons exhibit some complex vocalizations, but also because they depend on powerful arm movements to swing through the tree canopy.[38] Most gibbons actually walk bipedally about 6 to 7 percent of the time, which seems to reflect their highly specialized forelimb functions.[39] The pattern suggests a possible link between human bipedalism and carrying objects, on the one hand, and the lower position of the larynx, on the other.

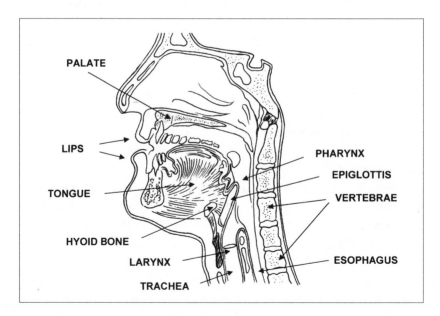

Figure 3.4 Unlike that of the hand, reconstruction of the vocal tract in fossil forms of *Homo* has been a major challenge in paleoanthropology, due primarily to the dearth of preservable skeletal parts associated with this organ.

Although many paleoanthropologists believe that language is closely linked to the evolution of the anatomically modern larynx, the relationship between them is somewhat ambiguous. A syntactic language is theoretically possible without a modern human vocal tract (or, for that matter, without *any* vocal tract [for example, sign language]). And while chimpanzees cannot produce the full range of speech sounds of which modern humans are capable (although some nonprimates, such as parrots, can imitate human voices with remarkable precision), no one language employs the full range of possible sounds and a syntactic language based on a much smaller range of sounds is conceivable.[40]

It is important not to lose sight of the extraordinary nature of human language.[41] In the context of evolutionary biology, it is a bizarre trait. Its closest parallel is the iconic honeybee dance, and it has no analogues among living vertebrates. Language often is studied as a vocal communication system that is found among many birds and mammals, but this recalls the study of Acheulean bifaces as stone tools. Just as the bifaces provide insights into the proto-mind of the Lower Paleolithic, language offers a key to the workings of the super-brain and the modern mind.[42]

The human vocal tract has an analogue in the hand. Like the hand, the vocal tract is a highly precise instrument that is capable of projecting very complex structures of thought outside the brain, using a range of sounds rather than a repertoire of movements. Through the conventions of syntactic language, the structures are in coded sequences composed of the sounds and entirely in digital and symbolic form, like the genetic code.[43] The range of sounds that may be generated by the modern human vocal tract is such that only a subset is needed to provide the digital elements of any given language.

But tracking the evolution of the vocal tract in the fossil record is much more difficult than reconstructing the changes in the morphology and function of the human hand. The key elements of the vocal tract are composed of soft parts or cartilage, and thus they do not preserve in the fossil record. Various researchers have attempted to reconstruct the vocal tract morphology of premodern humans on the basis of related skeletal parts that have been preserved. The degree of flexure at the base of the cranium was one suggested indication of a lowered position of the larynx, but the relationship between the two has been questioned. Other proposed clues have included the length of the neck and the size of the hypoglossal canal, but these also have proved problematic.[44]

At this point, there seems to be consensus that the modern human vocal tract had evolved by the time anatomically modern humans were present in Africa (that is, more than 100,000 years ago). And despite earlier efforts to demonstrate that the Neanderthals did not possess a modern vocal tract, it now appears that their vocal tract may have been similar to that of modern humans.[45] (Ongoing reconstruction of the Neanderthal genome reports the presence of FOXP2, a gene related to speech function and previously thought to be unique to modern humans.)[46] All this may indicate that anatomical speech capabilities antedate any archaeological evidence for behavioral modernity and, by implication, language.

The Levallois School of Language

The origin of language . . . seems to have been closely linked with technical motor function. Indeed the link is so close that employing as they do the same pathways in the brain, the two . . . could be attributed to one and same phenomenon.

ANDRÉ LEROI-GOURHAN

It is the archaeological record—with all its gaps and biases—that offers the trail of evidence for the evolution of the super-brain and the emergence of the modern mind. The broad pattern is this: roughly 300,000 years ago, some changes in the making of stone tools become apparent. They concern the way in which stone cores are prepared for the removal of flakes and seem to be tied to the production of composite implements (that is, weapons and tools composed of several parts). Both prepared-core, or *Levallois*, techniques and composite implements reflect an increase in the number of hierarchical levels and embedded components employed in creating artifacts.[47] And these changes seem to be tied to some other potentially significant developments, including the appearance of regional variations in the design of stone points.

If the changes seem less than dramatic, they nevertheless provide a logical prelude to the development of syntactic language, with its complex hierarchical structure and possibility for infinite creation. The premise is that the cognitive capabilities manifest in making more-complex hierarchically organized artifacts were eventually expressed in other media. The core property of modernity—potentially infinite combinatorial creativity—was expressed by making sentences and artifacts in the same way.[48] And if language itself is not the core property, its advent marks the threshold of modernity. Not only does it represent the primary medium by which modern humans rearrange the world around them, but it is the basis of the super-brain.

The process by which this threshold was attained remains unclear, how-ever. As with the expansion in brain size, no one seems to know what trig-gered the changes in stone technology 300,000 years ago, even though—in contrast to the large bifaces of the Lower Paleolithic—the practical benefits of the new technology are readily apparent. It is not clear if the process of development toward more complex forms took place gradually or in stages initiated by rapid bursts of change, or a combination of both. While prepared-core techniques exhibit progressive development over time, other changes appear abruptly.

It strikes me as plausible, if not likely, that feedback between individual brains and the increasingly complex artifacts, the externalized mental rep-resentations, played an important role in the process,[49] as did dynamic interactions among the brains within each social network. In chapter 2, I suggested that a similar process of feedback was critical to the development of the first externalized mental representations, in the form of bifaces. The role of speech, "artifacts" fashioned by the vocal tract in its presymbolic mode, may have been equally important but difficult to assess because of the uncertainties surrounding the evolution of the vocal tract.[50]

In its developed form, prepared- or Levallois-core technology is a set of procedures for generating flakes or blades of specific size and shape from large pieces of stone. Prepared-core techniques allow the toolmaker tighter control over the form of stone blanks struck off the core, and hence their logical connection to the production of composite implements, which require blanks of predetermined size and shape that can be inserted into wooden handles and shafts. Several steps are involved. First, the toolmaker must acquire a piece of stone of sufficient size and quality. Then he or she prepares a continuous surface (or series of "striking platforms"), extending around the perimeter of the stone, which is normally produced by succes-sive vertical blows to the upper face. The blows are struck with a "hard hammer" (another piece of stone). The upper surface of the core is shaped, in turn, by a series of horizontal blows around the perimeter. At this point, the toolmaker may choose from a variety of strategies for generating flakes or blades from the upper surface.[51]

The archaeologist Eric Boëda emphasized the hierarchical character of prepared-core technology (figure 3.5). Describing the relationship between the upper surface of the core and the platform surfaces around the perim-eter, he noted that "the two surfaces are hierarchically related: one pro-duces defined and varied blanks that are predetermined, and the other is conceived of as a surface of striking platforms for the production of

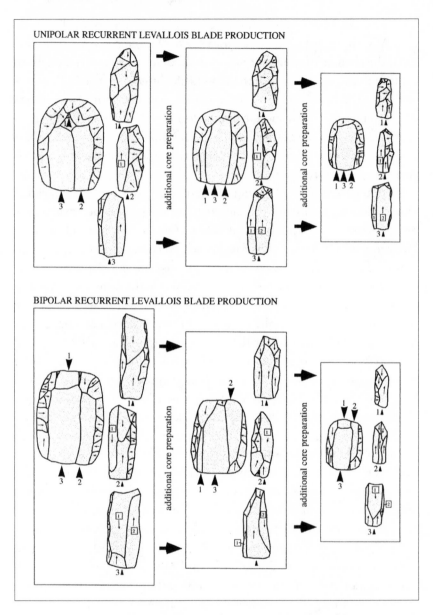

Figure 3.5 The dawn of creativity? Nonrandom variations in the reduction sequences of prepared-core technology, as reconstructed by Eric Boëda. (From Richard G. Klein, *The Human Career: Human Biological and Cultural Origins*, 3rd ed. [Chicago: University of Chicago Press, 2009], 488, fig. 6.33, redrawn from Eric Boëda, "Levallois: A Volumetric Construction, Methods, a Technique," in *The Definition and Interpretation of Levallois Technology*, ed. Harold L. Dibble and Ofer Bar-Yosef [Madison, Wis.: Prehistory Press, 1995], 59, fig. 4.28. Reprinted with the permission of the University of Chicago Press)

predetermined blanks. In the course of a single production sequence of predetermined blanks, the role of the two planes cannot be reversed."[52] Boëda also pointed to the collective character of prepared-core methods, which were shared among members of a group and "transmitted from generation to generation."[53] Over great distances of space and time, these groups devised nonrandom variations on prepared-core flaking strategies.

The origins of prepared-core technology seem to lie in the production of Acheulean bifaces, just as they probably may be traced to the bifacial flaking of the Oldowan pebbles. In some places, however, the relationship may be an indirect one. In Africa, prepared-core technology is said to have evolved from simpler Acheulean flaking techniques beginning around 500,000 years ago. But the simpler cores often were used to produce pre-forms for bifaces—that is, large flakes subsequently chipped into bifaces.[54] A fully developed prepared-core technology is present by the African Middle Stone Age, which began between 300,000 and 200,000 years ago.[55]

Prepared-core techniques emerged only among the makers of bifaces and their descendents, a pattern that is potentially significant. Fully developed Levallois cores appeared in western Eurasia more than 200,000 years ago, but never outside the *Homo heidelbergensis* range. Prepared-core technology was long believed to have been introduced into Europe rather suddenly, perhaps by the arrival of a large-brained *Homo* population from Africa.[56] Many now think that it developed independently from earlier biface technology in Europe.[57] In any case, the Neanderthals became skilled makers of both prepared cores and composite implements.[58]

The discovery that humans were assembling composite tools and weapons 250,000 years ago represents a major development in Paleolithic archaeology. Not one composite artifact has yet been recovered from the African Middle Stone Age or the Eurasian Middle Paleolithic, but their presence is established beyond a reasonable doubt from multiple converging lines of evidence. Composite implements are known among recent foraging peoples in many parts of the world,[59] and specimens have been recovered from sites that date to as early as the later Upper Paleolithic.[60]

In the Middle Stone Age and Middle Paleolithic archaeological record, composite tools and weapons can be reconstructed from traces of microscopic wear and adhesives on the stone blades and points that were inserted into wooden handles and shafts. In some cases, the microwear indicates the use of a binding cord or thong.[61] Microscopic traces of adhesives—including resin, bitumen, and ocher—have been found on the stone blades and points in both African and Eurasian sites.[62] Recent analysis of impact scars on blades from several South African sites indicates that they were attached in two different ways: on the end and on the side of a wooden shaft.[63]

The implications for human cognition are considerable, especially if the complex prepared-core technology is incorporated into the production process as one of several subassembly units (or "techno-units") that nest within a larger hierarchically organized structure.[64] The other techno-units require shaping and grooving a wooden handle or shaft and preparing and applying the binding agent(s). Each component subassembly involves obtaining and processing a different type of material. The components are brought together in a preconceived design that exhibits at least a few alternative forms—for example, stone-tipped spear versus hafted cutting tool or side-blade versus end-blade (figure 3.6)—and the parallels with language were noted by Stanley Ambrose:

> Conjunctive technologies are hierarchical and involve nonrepetitive fine hand motor control to fit components to each other. Assembling techno-units in different configurations produces functionally different tools. This is formally analogous to grammatical language, because hierarchical assemblies of sounds produce meaningful phrases and sentences and changing word order changes meaning.[65]

The structure and variety of Levallois cores and composite implements would seem to reflect two significant developments in brain evolution. The first is the increase in information-storage capacity—specifically, the expansion of the capacity for *working memory*, which has been defined as "task-relevant information" (such as memorizing directions to a particular address). New brain-imaging studies (functional magnetic resonance imaging [fMRI]) reveal that working memory is concentrated in the prefrontal cortex.[66] The second development is increased connectivity, which logically underlies the generativity manifest in the design variations of cores and composite implements. Here, too, recent fMRI analysis points to the prefrontal cortex. Both expanded working-memory capacity and increased generativity must have been essential to the development of language.[67]

Moreover, composite implements conferred obvious advantages on the people who invested time and materials on their production. If the practical functions of the Acheulean bifaces remain obscure, stone-tipped spears and hafted cutting and scraping tools must have been a major improvement over wooden spears and hand-held flakes in terms of power and

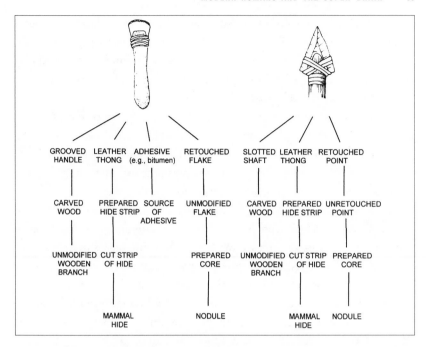

Figure 3.6 The production of composite tools and weapons required a sequence of steps for each "techno-unit" (representing a different raw material) within a larger hierarchically structured assembly and included at least several variants: scraper (*left*) and spear (*right*).

efficiency.[68] Perhaps other, unpreserved, technologies—also products of the emerging ability of humans to create more complex implements—contributed further to gathering food, defending against predators, and other aspects of human ecology. Their impact on long-term reproductive fitness may have been the critical evolutionary driver (directional selection) for more hierarchically organized technologies. And there is at least some support for this notion in evidence for changes in diet. In Europe, the Neanderthals display an impressive ability to bring down very large and undoubtedly dangerous animals, such as mammoth and woolly rhinoceros.[69] In Africa, there is an eventual broadening of the economy to include freshwater fishing and gathering of marine invertebrates.[70]

A more puzzling development during the period between 300,000 and 100,000 years ago is the use of mineral pigments. The pigments, which include fragments of red ocher (iron oxide) and black manganese dioxide, are found in both European and African sites. Some fragments exhibit wear facets that suggest application to a hard surface, and a few of them are described as "crayons"—that is, *drawing implements*.[71] Because mineral pigments were later used by modern humans to create drawings and paintings, some archaeologists suggest that they reflect a similar pattern among Neanderthals and modern humans of the African Middle Stone Age. If so, the artworks remain to be discovered.

An alternative explanation is that the pigments were part of the increasingly complex technology of the period. Recent foraging peoples used mineral pigment as an adhesive, or mixed it with other substances to produce an adhesive, for hafting stone blades or points to wooden shafts.[72] Now there is evidence that red ocher was used in a similar way during the African Middle Stone Age. Microscopic traces of ocher and plant resin were found on stone artifacts from Middle Stone Age sites in South Africa, indicating that a mixture of the two was applied to stone pieces to help secure them to wooden hafts.[73] This reveals yet more complexity in composite-implement production, adding further steps and another raw material to the preparation of the adhesive techno-unit. Lyn Wadley observed that the simultaneous appearance of mineral pigment and evidence for hafting in the archaeological record might not be a coincidence.[74]

The growing creative powers of the collective brain are manifest on a continental scale in the African Middle Stone Age. There is a pattern of marked regional variation in the design of stone artifacts, especially small bifacial forms and points, that apparently reflects emerging local styles and traditions (figure 3.7). Thus, for example, Middle Stone Age sites in North Africa yield distinctive stemmed points, while sites in west-central Africa contain equally distinctive narrow elongated bifacial points.[75] In a subdued way, the African record acquires a modern look by 250,000 years ago that is quite unlike the "random walk" pattern of the Lower Paleolithic.

Such a pattern of regional variation may be lacking in Europe, although some archaeologists have long insisted on the presence of Neanderthal "cultures." The famous French Paleolithic specialist François Bordes (1919-1981) proposed no fewer than five "tribes" in southwestern France on the basis of varying percentages of different types of Middle Paleolithic stone artifacts.[76] In the 1960s, Bordes's view was challenged by the American archaeologist Lewis Binford, who argued that functional differences and sampling error could account for the variations in artifact assemblages.[77] Others have suggested that perceived regional variations in assemblage composition elsewhere in Europe represent distinct local cultures. Here, too, it is difficult to discount the effects of functional differences

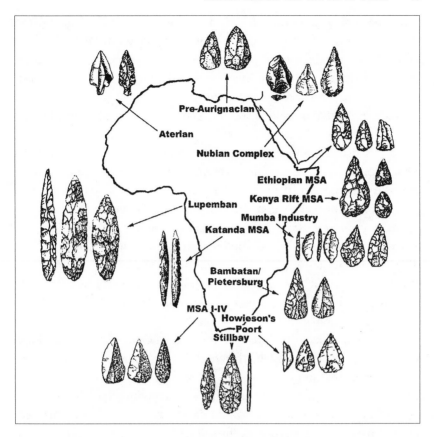

Figure 3.7 Regional variation in the design of small bifaces, points, and other stone implements in Africa during the Middle Stone Age yields a relatively modern-like pattern of
cultural variability. (After J. Desmond Clark, "African and Asian Perspectives on the Origin of Modern Humans," in *The Origin of Modern Humans and the Impact of Chronometric
Dating*, ed. M. J. Aitken, C. B. Stringer, and P. A. Mellars [Princeton, N.J.: Princeton University Press, 1993], 155, fig. 1. Reproduced with permission from the Royal Society, 1992)

and sampling error. Perhaps the sharpest contrast with the African record
is the absence of clear regional variations in the small bifaces and points.[78]

The cognitive archaeology of the Neanderthals may offer critical clues
to the emergence of the modern mind. As already noted, the Neanderthals had evolved brains of comparable size to those of modern humans
and may have possessed fully modern speech capabilities (if not syntactic
language). Independently or not, they developed prepared-core techniques
and composite implements, as did their counterparts in Africa. Although
some archaeologists maintain that the Neanderthals continued to progress

along the same lines as modern humans and eventually contributed to the rise of modernity,[79] most argue that they never got beyond the early stages of the change that began 300,000 years ago. Some variable or combination of variables implicated in the later stages of the formation of the modern mind seems to have been missing in northern Eurasia.

If the events that lie behind the emergence of language remain obscure, it seems to me that the combination of the archaeological record and existing knowledge of brain structure and function provide major clues about what happened. A basic premise—by no means accepted by everyone—is that the making of artifacts and the construction of sentences are fundamentally similar or exhibit some of the same properties.[80] The developments in prepared-core technology, composite implement design, and artifact style reflect increased combinatorial variation of elements, both discrete (digital) and analogical (continuous), on multiple hierarchically structured levels. At some point, a threshold was reached: the number of elements and levels was sufficient for generating a potentially infinite set of different combinations.

As Frederick Coolidge and Thomas Wynn noted, expanded working-memory capacity, localized in the frontal lobes, must have been a critical factor in reaching the threshold.[81] And presumably the role of the expanded prefrontal cortex in creating novel combinations also was critical. The problem is explaining how the capabilities of these and other neural structures connected to hand function also became connected to the vocal tract. Recent brain-imaging research does not provide an answer, but does reveal extensive overlap in areas activated for spoken language and symbolic "hand language" (specifically American Sign Language). Using PET scans, Karen Emmorey and her colleagues found that the left mesial temporal cortex and the left inferior frontal gyrus are "equally involved in both speech and sign production."[82]

This does not necessarily indicate that gestural language preceded spoken language (although it may have), but the finding strengthens the link between the function of the hands and that of the vocal tract for the externalization of mental representations, specifically those in symbolic mode. Based on the comparative functional anatomy and behavior of living primates, moreover, it is logical to assume that humans in Africa and elsewhere 300,000 to 250,000 years ago already relied on an extensive system of vocal communication.[83] The lower position of the larynx in humans relative to its position in other higher primates expanded the potential range of vocalized sounds, providing for increased combinatorial possibilities.[84] Viewed from this perspective, the emergence of spoken language seems less mysterious.

Documenting language in the archaeological record remains, nevertheless, an exasperating problem. If spoken language, as James Deetz observed, is a form of material culture shaped by the larynx rather than by the hand, and words and sentences are artifacts composed of sound, they are archaeologically invisible. But there are material forms of symbolism that survive. Until the development of writing, none of them represents language, but it is widely assumed that the making of symbols in material form with the hand probably indicates the production of symbols in the form of sentences with the larynx.[85]

While the evidence of increasingly complex, hierarchically organized technology in the form of prepared cores and composite implements after 300,000 years ago is unambiguous and indisputable, the evidence for symbolism before 75,000 years ago is highly problematic. The presence of mineral pigments that may have been used to draw some form of symbol, including simply a color without form, in Middle Stone Age and Middle Paleolithic sites has been noted. There also are isolated examples of mineral pigment in Lower Paleolithic sites dating to more than 500,000 years ago. In all cases, however, it is impossible to demonstrate that the pigments were used for any purposes other than purely technological.

At least two items recovered from west Eurasian sites, dating to between 500,000 and 250,000 years ago, represent possible examples of symbolism. A fragment of elephant bone from Bilzingsleben (eastern Germany) exhibits a sequence of evenly spaced, subparallel incisions that suggest an abstract design or a notation (for example, recording a series of objects or events). Similar patterns are often found on bone, antler, and ivory fragments after 40,000 years ago. The incised lines on the Bilzingsleben piece may have been caused by the cutting of soft material against the bone surface by a human using a sharp stone flake or the trampling of large animals, which is known to cause this sort of damage to bone.[86] Although neither seems likely to produce such evenly spaced lines of similar length, they cannot be categorically excluded as possible causes.

More intriguing is a piece of modified volcanic rock from the Levant that resembles a very simple and crude human figurine (figure 3.8). The site (Berekhat Ram) is located on the Golan Heights (Syria–Israel border) and is dated to between 420,000 and 350,000 years ago. The rock was incised in several places with another stone in what appears to be a nonrandom fashion. Deep grooves at one end of the rock create what might

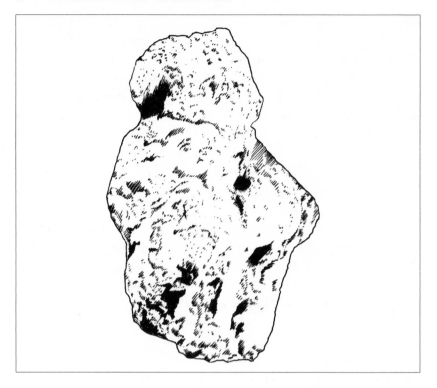

Figure 3.8 A piece of modified volcanic rock from Berekhat Ram on the Golan Heights may be a very early example of iconic (?) symbolism. (Drawn from a photograph by Ian T. Hoffecker)

represent the head of a figurine, while shallow parallel incisions on each side could delineate the arms. The Berekhat Ram object has been subjected to careful analysis, and there seems little doubt that the incisions were made by a human.[87] If it represents a piece of technology, its function is unknown. It is a fascinating, if isolated, possible example of early symbolism, apparently in iconic form.

The earliest well-documented evidence of symbolism may be personal ornaments in the form of perforated shells that show up roughly 135,000 to 100,000 years ago in the Near East and North Africa (figure 3.9). The Middle Paleolithic sites of Skhul (Israel) and Oued Djebbana (Algeria) yielded shells of the marine gastropod *Nassarius gibbosulus* that had been transported some distance from the seashore. Shells at Skhul had been perforated with a sharp object that was then rotated to enlarge the holes, presumably so that they could be suspended around the neck on some form of cord.[88]

Figure 3.9 Simple ornaments in the form of perforated marine shells appear in the archaeological record at least 135,000 years ago. (From Marian Vanhaeran et al., "Middle Paleolithic Shell Beads in Israel and Algeria," *Science* 312 [2006]: 1786, fig. 1. Copyright AAAS 2006)

It is not absolutely clear that the shells were used as symbols. Many animals have evolved various forms of ornamentation that is devoid of semantic or symbolic significance. Perhaps the shells were equally devoid of meaning, unlike the symbolic ornaments worn by people in historic times and today (for example, the peace sign). However, the subsequent pattern of ornament use, including the widespread utilization of perforated shells in the early Upper Paleolithic, suggests that they were worn to indicate social identity or group affiliation.[89] A pendant or necklace of marine shells would seem to have an entirely arbitrary relationship to social or ethnic identity and thus would constitute a true symbol. It may be significant that the earliest likely symbols are directly related to social interaction and, therefore, suggest integration of thought and the super-brain.

It also may be significant that shell ornaments are associated with the remains of anatomically modern humans at Skhul (reflecting a brief and limited expansion out of Africa roughly 100,000 years ago that preceded the main event, which occurred 60,000 years ago). Oued Djebbana is assumed to have been occupied by modern humans, since no Neanderthals are known from Africa. Unlike prepared cores, composite implements, and mineral pigment, perforated shells and other personal ornaments are absent in Neanderthal sites until the arrival of modern humans.[90] Their appearance in the archaeological record seems to mark a parting of the ways—the point at which modern humans appear to be moving inexorably toward modernity while Neanderthals uphold the status quo.

At Blombos Cave (South Africa), modern humans continued to produce personal ornaments from marine shells roughly 73,000 years ago, but also made some new types of tools: awls and polished points of bone. Most significantly, they incised geometric patterns on lumps of red ocher that appear to represent abstract designs.[91] Most archaeologists seem to regard the combined presence of these items, especially the symbols carved into the red ocher, as a sure sign that modernity had arrived.

The Archaeology of Alternative Reality

[Y]ou don't necessarily need the prefrontal cortex to form the mental image of a human or of a fish, but you need it to form the mental representation of a mermaid.

ELKHONON GOLDBERG

In terms of population ecology, the consequences of the super-brain were apocalyptic. Within a few millennia, modern humans spread across most of Eurasia and colonized Australia; only special circumstances—a wall of glacial ice—delayed settlement of the Western Hemisphere. They adapted with amazing speed to an extraordinary range of habitats and climate zones, from desert margins and wooded steppe to rain forest and shrub tundra.

Unlike earlier humans, they did not adapt genetically and differentiate into multiple species such as Asian *Homo erectus* and the European Neanderthals. Instead, modern humans used their creative powers and their ability to externalize their thoughts with hand and voice to generate alternative realities. They redesigned themselves as organisms and, in some places, redesigned parts of the environment. In cold regions, they fashioned fur clothing, complete with boots and hoods. To enhance their brain functions, they made artificial memory devices. In dark caves, they provided light with portable lamps. In landscapes devoid of caves, they built artificial shelters. To traverse large bodies of water, they assembled boats.[92]

In contrast to the relatively stable culture of the Middle Paleolithic and even the African Middle Stone Age, the archaeological record for this period, which is traditionally classified as the early phase of the Upper Paleolithic (or Later Stone Age), reflects continual and often dramatic change. There is a shift to Levallois stone-blade technology in the Near East, followed by a move to the mass-production of small stone blades.

Personal ornaments are common in sites of this period, along with a few bone implements.[93]

The appearance of people in Australia 50,000 years ago indicates watercraft. Other potentially complex technological innovations are suggested by the abundance of small-mammal remains in some Near Eastern and European sites, but the devices used to harvest these taxa (for example, snares, traps, and throwing darts) have yet to be retrieved. Eyed needles recovered from north Eurasian sites dated to 40,000 years ago indicate sewn clothing, which probably were made from many hierarchically structured parts and may have included mechanical components (for example, drawstrings for hoods). Both rotary drills and digging implements were present on the East European Plain by about 45,000 years ago.[94]

Visual art in the form of sculpture and cave painting illustrates creativity in complex hierarchical form. Each work of art is unique and reflects a potentially unlimited array of combinatorial variations. The *Löwenmensch* (lion man) ivory sculpture from Hohlenstein-Stadel (southern Germany) dates to about 35,000 calibrated radiocarbon years ago. Almost 1 foot high, it represents a human figure with the head of what appears to be a lion. The sculpture may be deconstructed into many components and embedded subcomponents, such as the head with multiple facial features or an arm with a hand and digits (figure 3.10).[95] The two-dimensional cave paintings from southwestern Europe, at least some of which date to the early Upper Paleolithic, exhibit even more complexity and variation.

Musical instruments also are found from this time period. The subtle shaping of pipes from animal bone is another impressive feat of early Upper Paleolithic technology as well as of accumulating technological knowledge (in this case, regarding the production of artificial sound variations).[96] In contrast to the visual art, however, the complexity of the generated musical structures remains unknown. The compositions may have been relatively simple. I am inclined to think that the complex process of making these instruments suggests otherwise, but there is no way to test this notion.

As during more recent times, the pace of change seems to have accelerated as modern humans continued to create alternative realities during the middle and later phases of the Upper Paleolithic. In some places, the size and complexity of settlements increased considerably as people invented new ways to harvest foods (for example, nets, spear-throwers [atlatls], and barbed harpoons) as well as to prepare and store them (for example, baking ovens and refrigerated pits). They figured out how to bake clay into ceramic objects, weave fibers into textiles, and domesticate dogs.

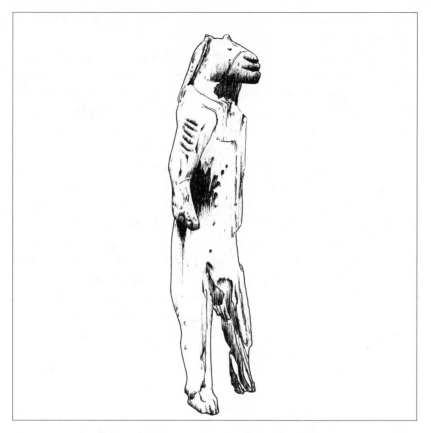

Figure 3.10 The *Löwenmensch* figurine carved from mammoth ivory, which was recovered from Hohlenstein-Stadel (southern Germany) and is dated to about 35,000 years ago, depicts an alternative reality: an imagined combination of human and lion. It is a complex representation with a hierarchical structure.

In the final millennia of the Upper Paleolithic, settlements in some areas acquired a village-like appearance that seems to anticipate postglacial sedentism and agriculture.[97]

History began with the emergence of the modern mind. From this point onward, the super-brain or collective mind—which reproduced itself in the form of various cultures as modern humans dispersed across the Earth—developed in its own way, as life had evolved after 3.5 billion years ago. By creating novel technologies and subsequently expanding and improving on them, it was accumulating knowledge about the external world. This knowledge was recorded and interpreted in the symbolic internal language of each culture or collective mind. The dynamism was, I believe, a product of the unstable and awkward relationship that existed then and now between the super-brain and the disgruntled organisms that generated and participated in it. The fossil record of the mind is the historical record, although it embraces more than the written records of the last few thousand years—it is represented by all forms of external thought.

4

THE UPPER PALEOLITHIC AS HISTORY

Our history has always . . . been up to us.

BARACK OBAMA

MORE THAN ANYONE ELSE, V. Gordon Childe (1892–1957) helped shift the subject of archaeology from artifacts to people. Nineteenth-century archaeology had been dominated by the classification of artifacts, organized into stratified sequences and interpreted within a framework based on a vague theory of progress. Beginning in 1925 with the publication of *The Dawn of European Civilization*, Childe transformed the artifacts into a historical narrative: "a preliterate substitute for the conventional politico-military history with cultures, instead of statesmen, as actors, and migrations in place of battles."[1]

Throughout his career, Childe searched for the appropriate framework to interpret the archaeological record from a historical perspective. In *The Dawn*, he applied the concept of *archaeological culture* to the artifacts, and it was this innovation that most impressed his colleagues in Britain. Childe had borrowed the concept from the German archaeologist Gustav Kossinna (1858–1931), whose views on the biological or racial basis of cultural progress enjoyed the public approval of leading Nazi officials.[2] Childe had rejected such ideas in his book, which instead emphasized the diffusion of ideas, and was distressed by their link to the archaeological culture concept. In the months after Adolf Hitler came to power in March 1933, Childe denounced the racial doctrines behind Kossinna's notion of culture in papers and lectures.[3]

In 1934, Childe revised an earlier synthesis on the Near East to include two major social and economic upheavals in prehistory: the rise of agriculture and the development of urban centers. He later referred to them as the *Neolithic Revolution* and *Urban Revolution*, respectively. They were modeled explicitly on the concept of the Industrial Revolution of the eighteenth and nineteenth centuries and, by analogy with the latter, were tied to a significant increase in population density.[4] Although Childe's thinking about the Neolithic and Urban revolutions was later attributed to the influence of Marxism and Soviet archaeology,[5] it was actually another step in his continuing effort to approach the prehistoric archaeological record from the perspective of a historian. In 1934, he shifted away from culture history and turned to the analysis of technological innovation, economic change, and their impact on population growth.

In the following year, Childe visited the Soviet Union, where he was inspired by the ongoing effort to apply a Marxist framework to the archaeological record. This framework was largely derived from the cultural evolutionists of the late nineteenth century, especially Lewis Henry Morgan. Although Marx and Engels had accepted it as the ethnology of their day, by the 1930s the unilinear evolutionary stages postulated by Morgan and his contemporaries had long been abandoned by ethnologists outside the Soviet Union. Nevertheless, Childe promptly published *Man Makes Himself* (1936), which combines his prehistoric economic revolutions with a Marxist viewpoint—simultaneously articulating his current thinking on prehistory and battling the racial archaeology of the Nazis.[6] He later wrote *What Happened in History* (1942), resurrecting the antiquated terminology of the nineteenth-century evolutionists by labeling the epochs preceding civilization as "savagery" (Paleolithic) and "barbarism" (Neolithic).[7]

In later years, colleagues wondered how serious Childe was about his commitment to Marxism and support for the Soviet Union; he often wielded them to shock people. Throughout his life, he was a shy and isolated person who never married or formed intimate friendships; some thought that his reticence reflected his rather odd physical appearance. He had emigrated from Australia to England, where he remained an outsider. He subsequently accepted an academic post in Edinburgh, where he repeatedly scandalized the local regime and later published—as a "parting insult"—a Marxist prehistory of Scotland. He wore strange hats and exotic clothing. Arriving at an expensive hotel, he demanded to know why copies of the *Daily Worker* were not available to guests. In public lectures, he outraged his audience with quotations from Comrade Stalin.[8]

Childe is probably best remembered for his Marxist and cultural evolutionary views. He became a victim of his own success, and *Man Makes Himself* and *What Happened in History* remain his most widely known works. When cultural evolutionism was revived in North America in the mid-twentieth century, he was solemnly acknowledged for his earlier writings. But Childe continued to think about an appropriate framework for the archaeological record, and during the last decade of his life, he developed something rather different from his earlier perspective. His principal biographer, Bruce Trigger, observed that "this was a period of lonely innovation, much of which was too easily dismissed by contemporary British archaeologists as more examples of 'Gordon's naughtiness.' It was also a period when major shifts occurred in his thinking, which produced real or seeming contradictions in his writings . . . an important creative phase . . . that significantly altered and developed his earlier contributions."[9] Childe was influenced by the British philosopher and historian R. G. Collingwood (1889–1943), and his later publications reflect Collingwood's idealist perspective.[10]

Collingwood was himself an isolated and iconoclastic figure. He practiced philosophy, history, and historical archaeology during a period when natural science had almost completely overwhelmed the philosophy of history. This was due in part to the impact of Darwin's ideas, but had deeper roots in the Enlightenment. Two popular historical works of the time were Oswald Spengler's *Decline of the West* (1926–1928) and the first volumes of Arnold Toynbee's massive *Study of History* (1934), both of which treat societies and cultures as organisms that develop through a natural life cycle. Collingwood regarded these studies as manifestations of the "tyranny of natural science" over history, which he believed to be "a special and autonomous form of thought."[11] He placed the human mind at the center of history: "Unlike the natural scientist, the historian is not concerned with events as such at all. He is only concerned with *those events which are the outward expression of thoughts*, and is only concerned with these in so far as they express thoughts" (italics added).[12] Shortly after the posthumous publication of Collingwood's book *The Idea of History* (1946), Childe wrote *History* (1947), which Trigger described as a turning point, a small volume in which he melded Collingwood's vision of history with his own interpretation of Marxism, emphasizing the role of invention and its relationship to social and economic change.[13] History, he wrote, is fundamentally a "creative process,"[14] and several years later, he openly rejected the cultural evolutionary schemes of the nineteenth and twentieth centuries for their implicit "magic force that does the work of the concrete individual factors that shape the course of history."[15]

Childe's later thinking was articulated in *Society and Knowledge* (1956), published a year before his death.[16] In this most interesting and thoughtful of his many books, he presented a theory of progress based on the observation that human societies increase their control over the environment by creating new technologies. By improving their ability to manipulate the world, societies acquire *knowledge* about the way it works (that is, the "sensory hand"), which they tend to explain in terms of the technologies (for example, the mechanistic worldview that developed in Europe after 1200).[17] Childe stressed the social character of both invention and the accumulation of knowledge. But the central theme of the book is human creativity, which he defined as "not making something out of nothing, but refashioning what already is."[18] He had at last found the appropriate framework to interpret the archaeological record as a historical narrative.

By defining history in terms of thought—"all history is the history of thought"[19]—Collingwood made prehistory, or at least a portion of it, part of history; as a historical archaeologist specializing in Roman Britain, he wrote about reconstructing the past "from documents written and unwritten."[20] And as Childe incorporated Collingwood's ideas into his later writings, he repeatedly referred to artifacts as "concrete expressions and embodiments of human thoughts and ideas."[21] The goal of the archaeologist, he stated, is "to recapture the thoughts" expressed by the people who created the artifacts, and he noted that, by so doing, the archaeologist "becomes an historian."[22] He emphasized that the accumulation of knowledge—the history of *science*—had begun long before the invention of writing.[23]

In theory, history as defined by Collingwood and Childe can be extended back to the Lower Paleolithic, especially since the appearance of the Acheulean bifaces 1.6 million years ago provides archaeologists with an externalized mental representation. Before the emergence of the modern mind, however, the lack of creativity manifest in the archaeological record makes for a rather dull narrative. It was not until the advent of the super-brain and modernity—of potentially infinite creativity in both symbolic and nonsymbolic form—that human societies began to accumulate and interpret knowledge on the scale that Childe had in mind. The interval described in chapter 3, corresponding with the Middle Paleolithic in Eurasia and the Middle Stone Age in Africa, seems to lie somewhere in between. It contains evidence of innovation and accumulating knowledge, but until the later African Middle Stone Age, it is very limited.

The archaeological record of the first 40,000 years of the modern mind is traditionally classified as the *Upper Paleolithic* (Later Stone Age in Africa). Like the other subdivisions of the early archaeological record, the term "Upper Paleolithic" is itself a piece of fossilized thought. It was proposed in western Europe during the mid-nineteenth century, when all prehistory was viewed as an epoch of gradual progress.[24] As more sites were excavated during the final decades of the nineteenth and the early years of the twentieth century, some archaeologists began to see a major break in the record at the start of the Upper Paleolithic.[25] Soviet archaeologists inserted this break into the Marxist evolutionary scheme that Childe initially embraced and later abandoned.[26] But at the beginning of the twenty-first century, the original classificatory framework remains in place. Upper Paleolithic is used here as a label with the sole virtue of familiarity.

The reclassification of the Upper Paleolithic as *history* is more than a matter of semantics and formal definition. It reflects recognition not only that the process of accumulating knowledge about the external world through creative innovation was already under way, but that the technological knowledge acquired was substantive and significant. It was knowledge that had a major impact on the human population; Upper Paleolithic settlements reached unprecedented size and complexity, and overall population density almost certainly increased. Moreover, it seems to have provided the requisite base of knowledge for what followed after the Upper Paleolithic: agricultural villages and urban centers. There is reason to doubt that these developments would have taken place without the accumulation of technological knowledge during the preceding 40,000 years.[27]

Upper Paleolithic history also reflects recognition that the accumulation of knowledge is the essence of the historical process—with or without written records—and that "prehistoric" archaeological data can provide a significant part of the story. We do not know a single word spoken by the peoples of the Upper Paleolithic, but we do know a great deal about their thoughts. Artifacts that offer evidence of technological knowledge—tools, weapons, facilities, and structures—are filled with information derived from the collective mind of each Upper Paleolithic society. These modified pieces of organic and inorganic material are literally part of the mind, remnants of hierarchically structured thought that existed (and still exists!) outside individual human brains. To understand history, Collingwood asserted, the historian discerns the thoughts of the past "by rethinking them in his own mind."[28]

Moreover, changes in artifacts and features observed over time (other than stylistic changes) provide a record of how thought progressed over time and how knowledge of the external world and how it works developed during the Upper Paleolithic. By the later phases of the period, people in many places understood more about physics, chemistry, and biology than had their predecessors, as well as how to apply this knowledge to achieve desired ends. The archaeological record of the late Upper Paleolithic reveals that people thought differently than they had earlier—their minds had changed.

The Age of Art and Music

Is the conception of nature and of social relations which underlies
Greek imagination and therefore Greek [art] possible when there are
self-acting mules, railways, locomotives and electric telegraphs?
KARL MARX AND FREDERICK ENGELS

The most mysterious and fascinating period in the modern human past is the interval between 50,000 and 30,000 years ago,[29] which is often referred to as the early Upper Paleolithic (or, simply and affectionately, the EUP). Along with the final phase of the African Middle Stone Age, it represents the initial outing of the super-brain and the modern mind. It may have been uniquely primitive, relative to later periods, with respect to technology because the people of the EUP were just beginning to create the tools, weapons, and other gadgets that would eventually be commonplace among foraging societies (especially at higher latitudes).[30] Adding to the mystery of this period is that EUP sites are often small and hard to find, particularly in places where caves and rock shelters are absent.

If the technology is primitive, the visual art and musical instruments of the EUP are remarkably sophisticated. Visual art includes sculptures of humans, animals, and a mixture of both, as well as the earliest dated cave drawings and paintings. The creative character of these iconic representations, which exhibit a potentially infinite array of combinatorial variations within a complex hierarchical structure, was presented in chapter 3 as an unambiguous manifestation of the modern mind.[31] And although the structure of EUP music is unknown, the subtle design of the wind instruments reveals an impressive knowledge of musical sound (figure 4.1).[32]

What is striking about the art and music of the EUP is not their prominence and sophistication compared with those of the later phases of the Upper Paleolithic, but their contrast with the technology of the time. During the middle and late Upper Paleolithic, the technology became

Figure 4.1 The earliest evidence for the creation of musical sounds and structures is at least broadly coincident with the earliest evidence for the production of visual art and creative technology, and surprisingly sophisticated musical instruments ("pipes") are dated to more than 30,000 years ago, including this specimen, recovered from Geissenklösterle (southern Germany). (Redrawn from Francesco d'Errico et al., "Archaeological Evidence for the Emergence of Language, Symbolism, and Music—An Alternative Multidisciplinary Perspective," *Journal of World Prehistory* 17 [2003]: 41, fig. 11b)

considerably more complicated, while the art and music—at least the design of the musical instruments—appears to have remained fundamentally unchanged.[33] The gulf between the structural complexity of art and of technology became even more pronounced after the Upper Paleolithic, and it has been especially stark in the machine era (and although music did become more complex, this trend seems to have been related to developments in the technology of musical-sound production). Moreover, written records show that the structure of language also remained fundamentally unchanged in later periods of accelerating technological complexity. The contrast between art and language, on the one hand, and technology, on the other, reflects the fact that they represent different parts of the mind. Art and language are almost entirely internal to the super-brain or collective mind; they are the primary means of super-brain integration. Through the medium of visual art, individuals project artificial representations that bear a rather obvious relationship to the internal natural visual representations of the catarrhine brain. Language is projected through a medium as well—the human voice—but this masks its true character. Language is integrated with sensory cognition (auditory, visual, and tactile [in the case of braille]), but its core elements have no sound, color, or feel; they probably reside in the neural pathways of the neocortex itself. Technology is both internal and external to the mind. Its internal aspect is the sharing of artifacts or thoughts about artifacts among individuals or groups. Its external aspect lies at the interface of the mind and the world outside the mind, where the mind engages the world. Most human technology is designed with nongenetic information, and an artifact is analogous to an organism—the phenotypic expression of that information. By the beginning of the EUP, modern humans had evolved a mental capacity analogous to sex—potentially infinite recombination of the bits of information.

Gordon Childe was one of many thinkers who noted the apparent connection between the clocks and other machines that arose in western Europe after 1200 and the mechanistic worldview that accompanied the new technology. He sensed a link between technology and the way the world is interpreted through language and art, although neither he nor anyone else has been able to explain precisely how it works. It seems clear, nevertheless, that as the structural complexity of technology has increased—bringing with it revelations of the ways in which the world works—the world has been reinterpreted in new ways (even if the structural complexity of the systems employed to explain it have remained fundamentally unchanged). The worldview of the early civilizations was probably as different from that of the hunter-gatherer societies that preceded them as it was from that of western Europe after 1200.[34]

If the technology of the EUP was uniquely primitive in comparison with those of the later phases of the Upper Paleolithic and of hunter-gatherers of the post-Paleolithic era, did EUP people perceive and explain the external world in their own peculiar way? And, given the widely held assumption among historians of technology that worldview affects the capacity for innovation,[35] did the way in which the EUP mind interpreted the outside world influence the pace of technological change during this mysterious period? It is impossible to answer these questions because we can only guess at the content of the EUP worldview and the meaning of the few symbols yielded by its archaeological record. As we study the pattern of change in all areas of material culture between 50,000 and 30,000 years ago, however, I believe that we should keep these questions in mind.

Some archaeologists view the EUP as "transitional" or intermediate between the late Middle Paleolithic and the later phases of the Upper Paleolithic. This idea is related to a once widely held notion that living Europeans evolved directly from the local Neanderthals. It remains popular among those who believe that the Neanderthals contributed genetically and/or culturally to modern humans. But it has become more difficult to sustain in recent years in light of the discoveries of EUP sculptures and paintings, as well as the new analysis of musical instruments, all of which suggest that the people of the EUP had "the same cognitive and communication faculties" as living humans. There are still dramatic differences between the EUP and the later Upper Paleolithic, but they seem to be confined to technology and economics—the same sort of differences apparent in historic times that reflect the gradual accumulation of new inventions and their impact on population density and settlement size.

Another factor that contributes to the false impression of the EUP as a transitional stage is the profusion of archaeological sites in many parts of northern Eurasia that date to this interval and contain a mixture of typical Middle and Upper Paleolithic artifacts. They are assigned to industries like the *Uluzzian* in Italy and the *Szeletian* in Hungary and other parts of central Europe. The artifacts include some types known only in Upper Paleolithic (or late African Middle Stone Age) industries, such as shell ornaments and bone points, alongside classic Middle Paleolithic stone-tool forms, such as scrapers made on flakes and small bifaces. In some cases, the mixture is thought to reflect the adoption of Upper Paleolithic traits by local Neanderthal groups; in others, it is viewed as the gradual development of technology by modern humans slow to abandon the traditions of the past.[36]

A more parsimonious explanation of the Middle Paleolithic tools in these sites is that they represent simple expedient forms that were used for certain tasks, especially the butchery of large-mammal carcasses. Modern humans continued to make such tools long after the EUP, and they are particularly common at sites in North America where large mammals were killed and butchered. The source of the confusion probably lies in the inherent biases of the west European archaeological record, which is heavily skewed toward *habitation* sites in caves (where these tool types are present but not abundant). For historical reasons, the interpretive framework of western Europe was applied far beyond its boundaries. In eastern Europe, where caves are often absent, a wider range of sites are found; some of them are *kill-butchery* locations that predictably contain many Middle Paleolithic tool types. In short, the presence of such tools in EUP sites probably has little to do with the Middle Paleolithic.[37]

Ironically, some of the earliest known traces of the EUP outside Africa are from distant Australia. When modern humans began to expand out of Africa, they apparently moved eastward across tropical Eurasia, perhaps initially along the coast of the Indian Ocean (figure 4.2).[38] Archaeological evidence of the south Asian migration remains to be found, but both artifacts and human skeletal remains show up in Australia by 50,000 years ago. Their presence alone is significant, since even the shortest water crossing of 55 miles from Southeast Asia indi-

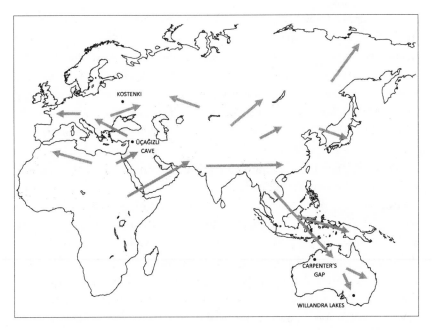

Figure 4.2 Modern humans spread out of Africa and into Eurasia and Australia roughly 50,000 years ago. Their rapid occupation of a wide variety of habitats and climate zones appears to have been largely a consequence of their unprecedented ability to redesign themselves with technology.

cates some form of watercraft, which seems to have precluded earlier settlement. Freshwater-shell middens are dated in both southeastern Australia (Willandra Lakes) and Papua New Guinea to more than 40,000 calibrated radiocarbon years ago, while evidence of marine-resource use is slightly younger.[39]

As in northern Eurasia, personal ornaments of perforated shell and other materials are found in the earliest phase of the EUP in Australia. And although musical instruments have yet to be recovered from an EUP context there, very early evidence of visual art is known at Carpenter's Gap (northwestern Australia) in the form of a fragment of a rock-shelter wall—apparently painted with ocher—that fell into deposits dated to roughly 46,000 to 39,000 years ago.[40] As elsewhere, technology remained relatively simple during the EUP in Australia. Points and other implements made of bone are generally scarce and thus far unknown from the earlier phases of the EUP (although this may be a function of sampling error and preservation bias).[41]

Above latitude 30° North, the EUP materializes on Africa's doorstep in the Levant also roughly 50,000 years ago. The archaeological record in the eastern Mediterranean is different from that in Australia and illustrates how, as they dispersed out of Africa into different parts of Eurasia and Australia, modern human groups began to develop their own local historical peculiarities. The succession of industries in the Levant between 50,000 and 40,000 years ago also illustrates the potential for rapid innovation and constant change.

Some isolated bone implements are found in this industry, as well as the ubiquitous personal ornaments in the form of perforated marine shells.[42] But the principal changes are manifest in the stone-tool technology, which began with a shift toward the production of long blades—in addition to triangular points—from prepared (Levallois) cores. As before, the stone points were attached to wooden shafts as spear tips, but the blades were chipped into tools that are considered generally typical of the Upper Paleolithic—especially end scrapers and burins. The technology gradually moved toward the mass-production of thin bladelets and narrow points that were struck off the cores with a "soft hammer" (bone or antler). The gradual character of the transition has been nicely documented in recent excavations at Üçağizli Cave on the southeastern coast of Turkey.[43]

The series of industries in the Levant after 50,000 years ago may illustrate something even more fundamental about the modern human mind than the capacity for rapid innovation. It is difficult to explain the changes, especially the later changes, as a result of external factors, such as climate or interactions with other species. The inhabitants of the Levant had begun to make their own history—driven by the internal dynamics of human society and the collective mind. While the initial changes in stone-tool technology could conceivably reflect the response of an incoming population from northeastern Africa to novel environments and/or local competitors (Neanderthals), it is hard to account for the subsequent shift to soft-hammer production of small blades and narrow points in either terms. The Neanderthals seem to have vanished from the Levant by 50,000 years ago, and the climate—although oscillating—was generally mild.[44]

Both climate change and nonhuman species have had effects on human societies throughout prehistory and history, but after the advent of modernity other factors have been at work. In historic times, following the emergence of the early civilizations, the primary external factor in shaping societies probably has been the impact of other societies. Climate change often has altered the course of history, such as the consequences of the Little Ice Age on European society and the impact of El Niño on the Classic Maya. Catastrophic events like the volcanic eruption on the island of Santorini roughly 3,600 years ago in the eastern Mediterranean also represent a periodic external factor. And other organisms, especially pathogens such as *Pasteurella pestis* (bubonic plague bacillus), have had an impact on history.[45] But there have been many developments, such as the invention of the printing press and the cotton gin, with significant consequences for human societies that cannot be plausibly ascribed to global warming, falling asteroids, or killer bees. These internal factors probably were at work during the EUP.

Most of what we know about the EUP is based on archaeological sites in Europe, which is a reflection of the long history of discovery and research in that part of the world. For the Upper Paleolithic, this is especially true in western Europe, where Paleolithic archaeology had its beginnings. It also reflects the high visibility of Upper Paleolithic sites in caves and rock shelters—nicely preserved and easily discovered—which are predominant in southwestern France and northern Spain but also known in Italy, Germany, and Austria. In central and eastern Europe, open-air sites are more common, and, in some areas, rock shelters are completely absent. Outside Europe, a number of Upper Paleolithic sites—including EUP sites—are known from southern Siberia.[46]

The northern bias of the EUP archaeological record has one advantage: it offers a rich source of data on how modern humans, recent migrants from the equatorial zone, adjusted to cooler and more seasonally variable environments. Their skeletal remains reveal anatomical proportions typical for living peoples of the tropics—tall and thin, with long limbs—who are adapted to warm climates.[47] Such people are more susceptible to cold injury and loss of core body temperature (hypothermia), and they require cultural buffering in the form of insulated clothing and heated shelters. One of the most striking aspects of the wide geographic dispersal of modern humans 50,000 years ago is how the biogeographic "rules" of climate adaptation did not apply. Human groups living in higher latitudes eventually developed stockier bodies and shorter extremities, which reduce heat loss and the potential for frostbite, but during the EUP, adaptation to northern environments was achieved almost entirely through the creation of new technologies.[48]

It may be significant that the earliest traces of the EUP in Europe coincide with a warm-climate oscillation. According to the Greenland ice cores, which provide the most detailed and comprehensive record of climate change in the Northern Hemisphere for the period (based on oxygen isotope fluctuations), temperatures increased sharply a few centuries after 48,000 years ago. Greenland Interstadial 12 began with a warm peak that was followed by a gradual cooling until colder climates returned shortly before 44,000 years ago. Artifact assemblages dating to this interval that bear a close resemblance to the oldest EUP industry of the Levant—that is Levallois blades and points with typical Upper Paleolithic tools—are found in southeastern and south-central Europe. Other than isolated bone implements, there is little evidence of novel technology, although this may be due largely to the biases of preservation, sampling, and site function. Most of these sites are thought to be stone-tool workshops.[49]

Because only isolated human skeletal remains that cannot be firmly identified as those of particular species are found in these sites, some archaeologists are hesitant to ascribe them to modern humans. In theory, the artifacts could have been made by local Neanderthals, and at least a few archaeologists are convinced that they were. Such a view is compatible with the widely held notion that the European Neanderthals produced several industries comprising a mixture of Middle and Upper Paleolithic artifact types.[50] However, there is consensus that the next batch of artifact assemblages to appear in southeastern Europe, at roughly 45,000 years ago, were most probably made by modern humans. The sites also lack diagnostic human skeletal remains, but they contain small bladelets and narrow points similar to those that were produced in the Levant before 40,000 years ago, which *are* associated with indisputably modern human remains in Lebanon, as well as other typical EUP artifact forms. Often referred to as the *Proto-Aurignacian* industry, this group of assemblages is found in caves and open-air sites in Italy and the Balkans.[51]

The EUP in Europe between 45,000 and 40,000 years ago is best represented on the East European Plain at a concentration of open-air sites on the Don River about 250 miles south of Moscow. The sites are located around a series of ravines incised into the west bank of the main valley around the villages of Kostenki and Borshchevo (figure 4.3). The EUP remains are buried in a thick bed of silt and fine rubble, inter-layered with lenses of carbonate formed by seeps and springs. The oldest levels lie beneath a volcanic-ash horizon deposited by an immense eruption in southern Italy (Campanian Ignimbrite) that spewed a cloud of ash across much of southeastern Europe. The ash is firmly dated to 40,000 years ago

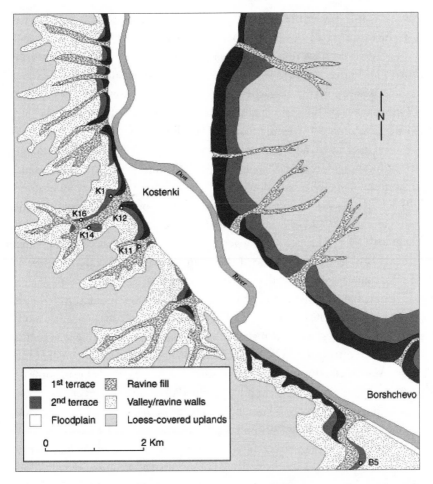

Figure 4.3 Much of the evidence for settlement on the East European Plain in the early Upper Paleolithic is derived from a group of sites around the villages of Kostenki and Borshchevo on the Don River (Russia).

and is represented in the Greenland ice cores, where it precedes a major cold-climate event.[52]

Mammoth bones were discovered at Kostenki centuries ago, but the associated Paleolithic artifacts were not recognized until 1879. In the 1930s, Soviet archaeologists excavated massive feature complexes of middle Upper Paleolithic age; the occupation floor patterns were used in the debates over the application of Marxist cultural evolutionary models to archaeology that intrigued Gordon Childe. Although older EUP occupations were known as early as 1928, they were investigated on a large scale only after

World War II, under the direction of A. N. Rogachev (1912–1981).[53] With-out the extensive earlier excavations of the younger levels, the EUP layers probably would have remained largely unknown because most are deeply buried and rarely exposed by erosion or other disturbances.

The excavations by Rogachev and more recent investigations exposed EUP occupation floors in many of the Kostenki sites (and lately at Bor-shchevo), ultimately revealing a unique "EUP landscape" across which peo-ple moved, camped, and engaged in various activities. Some of the sites rep-resent locations where large mammals—chiefly horse, but also reindeer and mammoth—were killed and butchered. They contain the same types of stone artifacts (points, bifaces, and scrapers) found at kill-butchery sites of different times and places. Other locations show traces of camps that were occupied for at least a few days, perhaps longer, yielding a diverse array of artifacts and other debris. The Kostenki–Borshchevo sites offer a more complete view of EUP society and economy than do the caves of southwestern Europe.[54]

The analysis of pollen-spore samples in layers below the oldest EUP levels at Kostenki indicates an interval of very mild climates that dates to more than 44,000 years ago and may correspond to Greenland Interstadial 12. Occupation during the EUP probably began at about the same time that the Proto-Aurigancian industry arrived in southern Europe, and at least some of same types of artifacts found in Italy and the Balkans are present at Kostenki–Borshchevo. In addition to bone points and awls, digging implements made of antler and ornaments perforated with a hand-operated rotary drill were found in the lowest levels. Perhaps most significant from a technological and economic perspective are the concentrations of small-mammal remains and some bird remains, indicating the expansion of diet and most probably the production of new devices for harvesting these foods (for example, snares, nets, and throwing darts). One of the Kostenki sites yielded what may be the oldest known visual art in the world—a small piece of carved mammoth ivory that resembles a human head and neck (figure 4.4). It may represent part of an unfinished human figurine that broke during carving.

Settlement at Kostenki–Borshchevo must have ended 40,000 years ago with the Campanian Ignimbrite eruption and ash fall across the East European Plain. In some places, the ash layer is several inches thick and presumably wiped out most plant and animal life for a while. The cata-strophic eruption also inaugurated a period of intense and sustained cold climate (known as Heinrich Event 4).[55] During this time, the most widely known EUP industry—the classic *Aurignacian*—spread across Europe. Con-ditions were so extreme in southwestern France that trees became scarce,

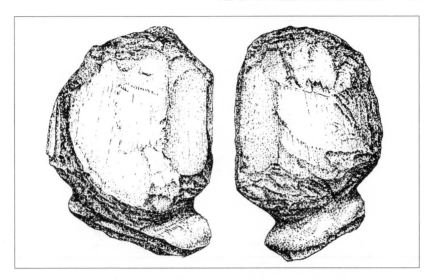

Figure 4.4 The earliest known specimen of visual art is a fragment of carved ivory from Russia that is dated to roughly 44,000 years ago and appears to represent the head and neck of an (unfinished?) human figurine (height = 2 inches). (Adapted from Andrei A. Sinitsyn, "Nizhnie Kul'turnye Sloi Kostenok 14 [Markina Gora] [Raskopi 1998–2001 gg.]," in *Kostenki v Kontekste Paleolita Evrazii*, ed. Andrei A. Sinitsyn, V. Ya. Sergin, and John F. Hoffecker [St. Petersburg: Russian Academy of Sciences, 2002], 230, fig. 9)

and the local inhabitants burned fresh bone as a substitute for wood fuel.[56] On the East European Plain, which eventually was reoccupied by Aurignacian folk, eyed needles of bone and ivory provide early evidence of sewn clothing (figure 4.5A).[57]

Overall, the technology of the later EUP remains simple in comparison with that of later periods, however, and the size and complexity of the settlements are modest. The best-known example of Aurignacian technology is the split-base point carved from a piece of deer antler, which was designed to wedge tightly into the socket of a spear shaft, reflecting familiarity with the elastic properties of antler and the way to exploit them for improved spear construction (figure 4.5B).[58] A fishing gorge from Italy indicates cleverly applied knowledge of fish behavior. Also noteworthy is the evidence for simple information technology, represented by fragments of bone and ivory that exhibit sequences of engraved marks. But many of the more complex technologies of the later Upper Paleolithic, including mechanical spear-throwers (atlatls) and high-temperature kilns, were still to come. The occupation areas of EUP sites are relatively small. Traces of

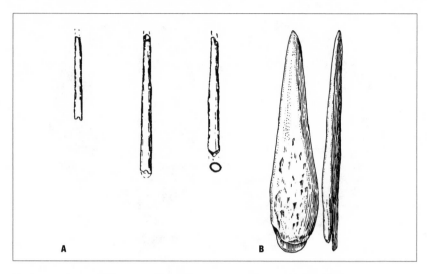

Figure 4.5 Two notable technological innovations of the EUP: (A) eyed needles manufactured from bone and ivory date to around 35,000 years ago in eastern Europe and presumably reflect the need for sewn winter clothing (after A. N. Rogachev and A. A. Sinitsyn, "Kostenki 15 [Gorodtsovskaya Stoyanka]," in *Paleolit Kostenkovsko-Borshchevskogo Raiona na Donu, 1879–1979*, ed. N. D. Praslov and A. N. Rogachev [Leningrad: Nauka, 1982], 170, fig. 59); (B) the split-base point was an application of the elastic properties of deer antler (redrawn from Paul Mellars, *The Neanderthal Legacy: An Archaeological Perspective from Western Europe* [Princeton, N.J.: Princeton University Press, 1996], 394, fig. 13.1).

former dwellings are rare and indicate ephemeral structures—tents erected with wooden poles and walls of mammal hide—apparently used for limited periods by small groups.

It is during this phase of the EUP that the contrast between the visual art and the musical instruments, which are represented by spectacular finds at sites in western Europe, becomes so striking. Controversy continues to surround the dating of the paintings in Chauvet Cave (southern France), which may be younger than previously reported, but they can be tentatively assigned to the EUP. In any case, the *Löwenmensch* sculpture from Hohlenstein-Stadel, as well as several recently discovered Aurignacian sculptures from Hohle Fels Cave (southern Germany), illustrate the capacity to create visual art of comparable complexity to that at Chavet. Among the objects from Hohle Fels are a female figurine and a therianthropic representation reminiscent of the *Löwenmensch*.[59] To these may be added the famous animal carvings from Vogelherd Cave (southern Germany) and the "dancing lady" stone sculpture from Galgenburg (Austria).

Finely crafted wind instruments (pipes) from Geissenklösterle (Germany) and Hohle Fels Cave are from the same time range.[60]

The later EUP occupations at Kostenki yielded at least one burial containing, in addition to the bones and teeth of the deceased, grave offerings in the form of ornaments and tools as well as generous inclusions of red and yellow ocher.[61] If the meaning of EUP visual art remains obscure, there can be little doubt that the grave offerings pertain to belief in an afterlife. The steadfast denial of the reality of death is one of the most characteristic elements of every known culture and reflects the ghastly dilemma that confronts an organism with the faculty of abstract thought or "intentionality"—recognition of the inevitability of its own death.[62] Only behaviorally modern humans seem to have faced this dilemma; there are some Neanderthal burials, but they lack any convincing traces of ritual and symbolism.[63]

The Gravettian: An Industrial Revolution

Technology is explained by history and in turn explains history. . . .
It is not a linear process.
FERNAND BRAUDEL

Gordon Childe may have overlooked the first economic revolution in prehistory, which arguably took place between 30,000 and 25,000 years ago in northern Eurasia. Perhaps he did not regard the social and demographic consequences as sufficiently profound, especially when measured against the tectonic upheavals of food production and urbanization.[64] It seems, nevertheless, to have followed the classic pattern that Childe envisioned: a wave of technological innovation associated with significant growth in settlement size and population density. But in an Upper Paleolithic context, it is difficult to sort out the cause-and-effect relationships. Many of the innovations seem to have been unrelated to economy and population and may have been a result instead of a cause of the increased size of communities.

In Europe, this phase of the Upper Paleolithic (often referenced as the middle Upper Paleolithic) is associated with the *Gravettian* industry or culture, which is broadly distributed across the continent from Spain to Russia. The Gravettian is best known for its evocative "Venus" figurines, carved in ivory and stone and sometimes modeled in fired clay. Another characteristic and common artifact is the shouldered stone point (often used as a knife). The most significant archaeological remains of the Gravettian, however, are the often large and complex settlements comprising hearths, pits, and other features.[65] These settlements suggest gatherings, at least on a temporary basis, of unprecedented size and are the principal basis for the inferred growth in population. Traces of similar settlements, although not classified as Gravettian, are found in southern Siberia at this time.[66]

The Gravettian emerged from the EUP immediately before the beginning of the Last Glacial Maximum (23,000 to 22,000 years ago). Climates were becoming increasingly cold and dry—especially in eastern Europe and Siberia, where isolation from the moderating influence of the North Atlantic Ocean produces more continental conditions. Periglacial steppe environments inhabited by a curious mixture of arctic-tundra and temperate-grassland species developed in these regions. Some archaeologists have suggested that the burst of technological innovation associated with the Gravettian was a response to the cooling climates.[67] There is a strong correlation between technological complexity and latitude among foraging peoples,[68] and at least some of the novel instruments and devices of the Gravettian reflect the presence of cold conditions.[69] But the pace of technological innovation continued, and probably even accelerated, after the Last Glacial Maximum and during the period of gradually warming temperatures that followed.

As in the EUP, I suspect, internal factors were at work, for external variables such as climate and other species probably account for only some of the culture change that took place during Gravettian times. The critical question is one that Childe raised in his best known books: What accelerates innovation, and what retards it?[70] What helps unleash the creativity of the mind—and facilitates the acceptance of new ideas—and what suppresses novel thinking? History is filled with examples of societies that stagnated, some of which suffered partial or complete destruction as a consequence; there are also historical examples of societies that experienced waves of invention and dramatic change.[71] Less attention has been paid to prehistoric societies in this regard, but it is apparent from the archaeological record that foraging peoples of the past underwent similar episodes of change, as well as protracted periods of stability. There may be a strong tendency among archaeologists to ascribe the advances in prehistoric societies to environmental factors, but I see no reason to exclude a role for the same variables, such as worldview and social institutions, that seem to have affected innovation in historical societies.

The population growth associated with the appearance of the Gravettian is inferred from the size and complexity of the remains of the settlements. The largest sites are thought to have been places where a large number of people (perhaps 50 to 100) assembled for at least a few days and perhaps several weeks. Although the famous Gravettian sites are in central Europe (for example, Dolní Věstonice [eastern Czech Republic]), the most extensive settlements seem to lie on the East European Plain at places like Avdeevo, Zaraisk, and the uppermost level at Kostenki 1, all in Russia (figure 4.6).[72] At these sites, linearly arranged hearths are surrounded by large pits; they apparently represent integrated-feature complexes constructed during a single episode of occupation. Perhaps some or all of these temporary aggregations were designed to exploit a concentration of a food resource, such as a fish run; stable isotope analyses of human bone from several sites indicates a high level of consumption of freshwater foods.[73] It seems likely, however, that these complexes were also used to reinforce social ties through rituals, feasts, and other public events.

Some of the population growth may reflect the impact of periglacial climates on flora and fauna in eastern Europe and southern Siberia. The cold and aridity were favorable to steppe species like bison and tundra dwellers like reindeer, and the large-mammal biomass may have increased relative to that in the generally milder conditions that had prevailed during the EUP. But a significant factor in population growth probably was the accumulated technological knowledge of the EUP—perhaps, especially, the late EUP—and possibly a wave of innovation that marked the beginning of the Gravettian. Novel inventions and improvements over existing technologies to harvest, prepare, and store food probably increased the number of people who could support themselves per unit area. Gravettian sites yielded the first evidence of nets, for example,[74] which may have been used to catch birds, fish, and small mammals like hare. A boomerang is reported from a site in Poland.[75]

The Gravettians knew how to manipulate temperature for various purposes, and they constructed miniature worlds in which temperature could be controlled. At least some of the large pits in their sites were apparently dug down to the permafrost level during warmer months to create "ice cellars" for cold storage of perishables (figure 4.7A). In areas where wood was scarce or absent and bone was used as fuel, these pits probably were used to keep supplies of bone fresh and flammable.[76] Even more impressive was their fired-ceramic technology. Modeled-clay objects were heated to temperatures of at least 1,500°F in constructed kilns (figure 4.7B).[77] Like

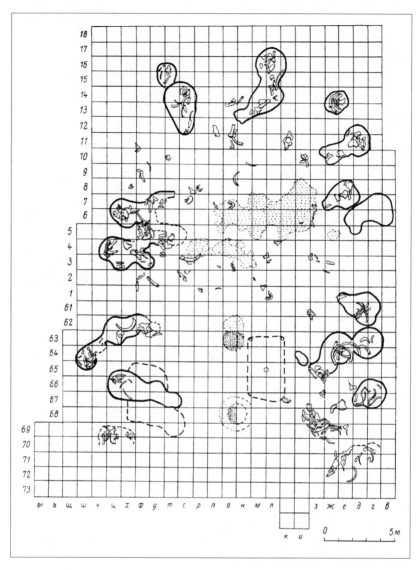

Figure 4.6 A Gravettian settlement in Russia, illustrating the complex arrangement of hearths and pits: Kostenki 1, Layer I, second feature complex, was excavated in the 1970s by A. N. Rogachev and others. (Modified from A. N. Rogachev et al., "Kostenki 1 [Stoyanka Polyakova]," in *Paleolit Kostenkovsko-Borshchevskogo Raiona na Donu, 1879–1979*, ed. N. D. Praslov and A. N. Rogachev [Leningrad: Nauka, 1982], 45, fig. 11)

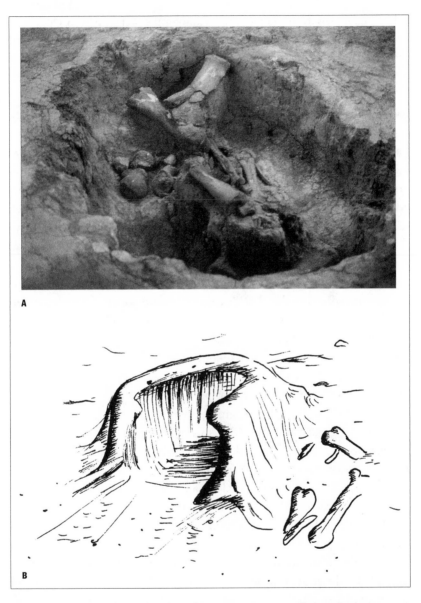

A

B

Figure 4.7 Technological applications of temperature in controlled artificial micro-environments: (A) a cold-storage pit at Kostenki 11 (Russia) was dug to what was probably the level of frozen ground during warmer months (photograph by author); (B) an excavated kiln at Dolní Věstonice I (eastern Czech Republic) could fire clay at temperatures of about 1,500°F (redrawn from Pamela P. Vandiver et al., "The Origins of Ceramic Technology at Dolni Věstonice, Czechoslovakia," *Science* 246 [1989]: 1007, fig. 7a).

other major new technologies—for example, the weight-driven clock—the production of fired ceramics among the Gravettians seems to have been devoted to nonutilitarian purposes, such as making art objects. The practical uses of this technology did not emerge until the later Upper Paleolithic.

The Gravettians manufactured a profusion of household items and gadgets, some of them of unknown function. They made portable lamps, fueled with animal fat, out of mammoth bone, and apparently they weaved baskets. As in EUP times, they made sewing needles from bone and ivory and fashioned small needle cases similar to those of the Inuit. The southern Siberian sites yielded figurines that depict people wearing fur suits, complete with snug-fitting hoods, which illustrate this highly complex technology for the first time. But the function of many of the carved and drilled pieces of bone and ivory in Gravettian sites is unknown; they are reminiscent of the items recovered from prehistoric Inuit sites.[78]

The increased population density in the Gravettian is significant because it represents the first discernible growth of the mind or super-brain. Although the expansion of modern humans out of Africa must have greatly increased the total number of individuals who contributed to the collective mind, the relative simplicity of EUP technology constrained "carrying capacity" and population density, as indicated by the consistently small size of the sites of that period. Day-to-day social interactions must have been limited to small groups, supplemented by periodic interactions among groups and individuals within wider networks (maintained by marriage and trade). As modern humans dispersed across Eurasia and Australia, they would have differentiated into local groups and regional networks—with corresponding dialects and languages—which are reflected as separate cultures in the archaeological record.

The large gatherings at Gravettian sites would have represented an expansion of the individual components of a super-brain in any given area or, at the very least, an intensification of interactions and collective thinking in a local neocortical network. As a larger gene pool offers a wider source of genetic variation, a super-brain composed of more components offers a greater potential for innovation and creativity. And as a bigger organism provides more room for hierarchically organized specialization among its cells and organs, a super-brain with more participants contains more potential for specialized thinking—for individuals devoted to specific arts or technologies. Although full-time specialists are not thought to have emerged until later times, with the advent of much larger settlements, the seeds were present in the kiln makers and tenders of the Gravettian.

The pattern of Gravettian art may illustrate another consequence of greater population density and an enlarged collective mind. An increased potential for creativity and innovation also brings an increased potential for alternative realities in the form of subversion and blasphemy. If it contains the seeds of specialized thought in some of its complex technologies, the Gravettian may also yield traces of another phenomenon that emerges vividly in later times and among more complex societies: the control of thought and behavior through public ritual and other expressions of faith.[79] Childe counted it among the forces—perhaps the most important one—that inhibit innovation.[80] The often elaborate burials of the Gravettians—such as the triple grave at Dolní Věstonice II, with its inclusions of ocher, charcoal, and various artifacts[81]—may reflect the intensification of public ritual, but it is the mobilary art that suggests a break with the EUP.

The "Venus" figurines of the Gravettian are striking for not only their subject matter, but also their ubiquity and geographic spread (figure 4.8).[82]

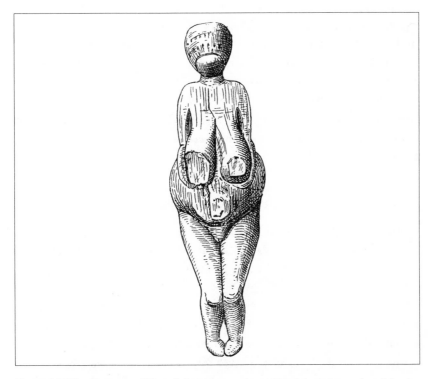

Figure 4.8 The Gravettian "Venus" figurines are the earliest known examples of the use of a uniform motif in visual art, although they exhibit some stylistic variation. The pattern might reflect changes in society and politics during this first "industrial revolution."

They contrast markedly with the EUP sculptures, which are few and highly individualistic. The female figurines may reflect the emergence of a Gravettian worldview, and while its content may forever remain unknown, the sameness of the sculptures across a vast expanse suggests widespread conformity to the new doctrines. Future discoveries of, say, more *Löwenmenschen* at EUP sites may expose the contrast as a sampling problem. But at present, a difference between the art of the EUP and that of the Gravettian is apparent, and it could be interpreted as evidence of a conservative backlash against the social and economic upheavals that accompanied the end of the EUP or simply as a response to the unprecedented cacophony of voices that arose with the Gravettian. Perhaps the wave of innovation came to a close at this point.

In any case, the Gravettian way of life seems to have ended abruptly after 24,000 years ago, and at this point, climate may indeed have been the cause. Both on the East European Plain and across southern Siberia, settlement declined markedly during the Last Glacial Maximum, and some stratified sites exhibit an occupation hiatus.[83] It is unclear why people appear to have abandoned the coldest and driest regions of mid-latitude Eurasia at this time. In addition to the extreme cold, food and fuel resources probably decreased, and one or more of these variables may have reached a critical threshold for human settlement. Skeletal remains of the Gravettians reveal that they retained in large measure the anatomical proportions of their EUP ancestors—an essentially tropical physique—which may have been a significant liability despite their fur clothing.[84]

The Periglacial Village

The archaeological course toward domestication in the Levant can be traced from around 19,000 B.C., at the peak of the last glaciation.
PETER BELLWOOD

Climates in the Northern Hemisphere ameliorated somewhat after the Last Glacial Maximum, but temperatures remained low for several millennia (19,000 to 16,000 years ago). Significantly milder conditions prevailed during the Lateglacial Interstadial (15,000 to 13,000 years ago), followed by a brief cold interval at the end of the Pleistocene epoch, or Ice Age (which terminated after 11,000 years ago). For the first time in more than 100,000 years, full interglacial climates returned to the Northern Hemisphere.

There is broad consensus among prehistorians that warming climates and their effects on plants and animals after 11,000 years ago were the primary cause of settled village life, farming and stockbreeding, and eventually urban centers and civilization. I suggest instead that climate change had a limited role in the development of villages and cities. A more significant factor was the steady accumulation of technological knowledge during the critical millennia between 19,000 and 11,000 years ago. I further suggest that village agriculture would not have developed anywhere, at least not for many thousands of years, had the Ice Age ended 19,000 years ago, before this important phase in the history of the mind.

Building on the existing knowledge of the early and middle Upper Paleolithic mind, societies in various parts of the world achieved major breakthroughs and expanded and refined previously known technologies. They developed the earliest known mechanical technology, applied their high-temperature kilns to the production of pottery, and domesticated dogs. Many of the new technologies hint at a more settled way of life, and sites in a number of regions acquired an increasingly village-like appearance—groups of houses with walls of bone in some cases, and paved stone floors in others.[85] In the Near East, where the oldest known agricultural settlements were discovered long ago, the gradual transition from hunting camps to farming villages—accompanied by significant population growth—began after the Last Glacial Maximum.[86]

Gordon Childe described the first artifacts with moving parts as "engines," even though human muscles provided the power source.[87] They constitute a defining difference, nevertheless, between the technology of humans and that of all other organisms. The oldest documented specimen is a fragmentary spear-thrower from Combe-Saunière I (southern France) that dates to about 20,000 years ago (figure 4.9).[88] As the earliest known mechanical technology, it is predictably simple and represents a classic example of "organ projection" envisioned by the nineteenth-century philosopher of technology Ernst Kapp,[89] by effectively extending the length of the arm and creating an artificial hand to increase its leverage. The bow and arrow appeared much later; wooden arrow shafts from Stellmoor (Germany) are dated to no more than 12,000 years ago.[90] While the design of the arrow was a modification of that of earlier projectiles, the bow was a novel and highly imaginative piece of technology without obvious parallels or models in the natural world—its structure was generated in the mind. Human muscles still provided the energy,

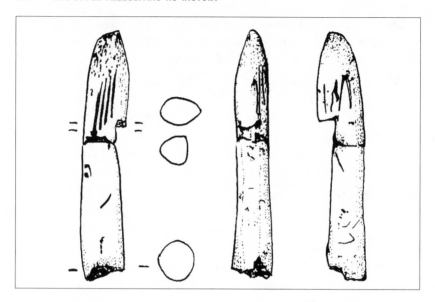

Figure 4.9 The oldest currently known example of mechanical technology is a fragment of a spear-thrower (atlatl) from Combe-Saunière I (southern France). (Redrawn from Pierre Cattelain, "Un crochet de propulseur solutréen de la grotte de Combe-Saunière 1 [Dordogne]," *Bulletin de la Société préhistorique française* 86 [1989]: 214, fig. 2)

but it was stored in the bow and bowstring by applying knowledge of the elastic properties of these materials.

More complex mechanical technology—also powered by stored energy supplied by muscles, but without a human present to release it—may be manifest in "untended facilities" for trapping small mammals. The harvesting of small mammals can be traced back to the initial phases of the EUP, but the technology behind it (perhaps nets and snares) remains unclear. Some possible trap components in the form of ivory fragments that were planed, beveled, and otherwise worked into various shapes were found at Mezhirich (Ukraine) (figure 4.10) dated to roughly 17,000 years ago.[91] An effective trap reflects knowledge of animal behavior and how to exploit it and is a clever device for saving both time and labor. Thomas Hobbes would have recognized it as the first example of "artificiall life,"[92] designed to function like a simple organism. Because traps would have been placed away from settlements—potentially distributed across the landscape along extensive traplines—they have unusually low archaeological visibility and may have been used much more often than an isolated specimen would suggest.

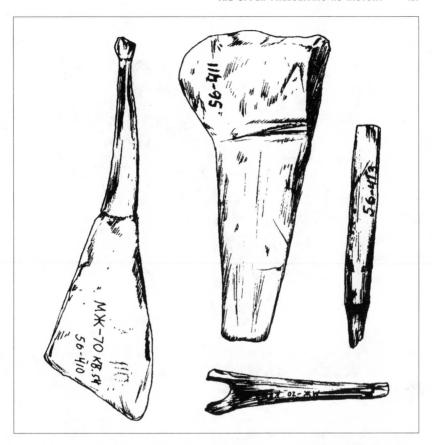

Figure 4.10 Several pieces of modified ivory recovered from Mezhirich (Ukraine) have been interpreted as trap components. Evidence for the consumption of small mammals and freshwater aquatic foods in earlier sites suggests that traps and snares probably date to the EUP. (Drawn by Ian T. Hoffecker, from a photograph in I. G. Pidoplichko, *Mezhirichskie zhilishcha iz Kostei Mamonta* [Kiev: Naukova dumka, 1976], 164, fig. 61)

Pottery was long associated with the rise of village farming, but it is now apparent that people were producing ceramic vessels in the immediate aftermath of the Last Glacial Maximum. A partially reconstructed pot—a wide-mouthed, cone-shaped vessel—from Yuchanyan Cave (southern China) dates to between 18,300 and 15,430 years ago (figure 4.11).[93] The dates are only slightly older than those for the earliest pottery found in Japan (roughly 16,000 calibrated radiocarbon years ago) and for late Upper Paleolithic pottery in the Amur River Basin (Russian Far East), at 16,500 to 14,500 years ago.[94] Some of the pots from Japan were manufactured with fiber temper and decorated with simple incised or impressed

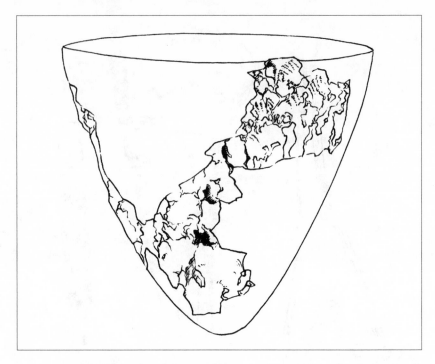

Figure 4.11 Reconstructed ceramic vessel from Yuchanyan Cave (southern China). (Drawn from a photograph by Ian T. Hoffecker)

designs. Traces of secondary burning and carbonized adhesions indicate that many were used for boiling and cooking, while others may have been used for storage.[95]

Domesticated dogs are documented at Eliseevichi I (Russia), in the form of skulls with shortened snouts, a diagnostic characteristic that helps differentiate them from wolves.[96] The skulls are associated with an occupation layer that contains remains of dwelling structures and substantial quantities of household debris, dating to roughly 17,000 years ago. More recent dogs are reported farther west at Bonn-Oberkassel (Germany).[97] An analysis of mitochondrial DNA among living dogs placed their initial domestication in East Asia during the later Upper Paleolithic,[98] which is not surprising given the hint of protracted settlement provided by the pottery. Dogs represent the first *biotechnology*—the genetic modification of an organism—even if undertaken with little sense of purpose or method.

Another realm of innovation laden with destiny is information technology. There is evidence for simple artificial-memory systems, comprising rows of incised marks on pieces of bone, in the EUP. More convincing examples have been recovered from Gravettian sites, such as Abri Labuttat (France), but the most complex forms, sometimes containing hundreds of marks, date to the later Upper Paleolithic and include specimens from Laugerie Basse and La Marche (both France) and Tossal de la Roca (Spain) (figure 4.12). These pieces reveal groups of marks—for example, short parallel strokes and notches—some which were made with different tools and perhaps at different times. A few of the marks on the Laugerie Basse piece were partly erased.[99] The information recorded, and apparently deleted, remains unknown. For decades, Alexander Marshack argued that at least some of them represent lunar calendars,[100] but others have questioned or disputed his conclusion. An alternative explanation is that they record the number of prey taken, and it is noteworthy that engraved images of two horses are associated with the marks on the La Marche piece.[101]

If tailored clothing is artificial skin and spear-throwers are elongated arms, the notational systems of the Upper Paleolithic are examples of a technological extension and enhancement of brain function outside the individual cranium. They also illustrate the use of true symbols—the shape of the marks presumably having little connection with the items recorded—in digital rather than iconic form. On some pieces, the marks seem to be organized hierarchically, and, in my view, they provide the strongest indirect evidence for the presence of syntactic language.

Perhaps the best evidence for increasingly sedentary life during the later Upper Paleolithic is offered by the former settlements themselves. If Yuchanyan Cave and other sites in East Asia that have yielded fragments of pottery are regarded as precursors to the village life of the post-Paleolithic era, large open sites in other regions contain the remains of village-like groups of semipermanent dwellings. The most widely known are the mammoth-bone houses of the East European Plain (figure 4.13A). In a landscape devoid of trees, the houses were constructed from the bones and tusks of woolly mammoth. Most are round or oval in plan and several yards in diameter. Former hearths (filled with bone ash) are located in the interior of each residence, along with a mass of household debris. Deep storage pits were dug around the houses, possibly for cold storage of bone or food during the warmer months. Traces of no fewer than four mammoth-bone structures have been found at Mezhirich and Dobranichevka (Ukraine) and Yudinovo (Russia) (figure 4.13B).[102]

Rather permanent-looking former dwellings also are known from late Upper Paleolithic open-air settlements in western Europe. In southwestern

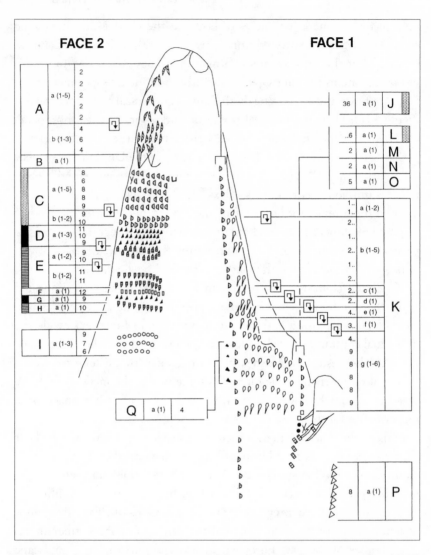

Figure 4.12 Information technology in the form of artificial memory systems (AMS) dates to the earlier phases of the Upper Paleolithic, but became increasingly complex during the later phases. Each face of this antler from La Marche shelter (France) is engraved with a series of marks interpreted by Francesco d'Errico as groups of signs. (From Francesco d'Errico et al., "Archaeological Evidence for the Emergence of Language, Symbolism, and Music—An Alternative Multidisciplinary Perspective," *Journal of World Prehistory* 17 [2003]: 34, fig. 8[f]. Reprinted with the permission of Springer Science and Business Media)

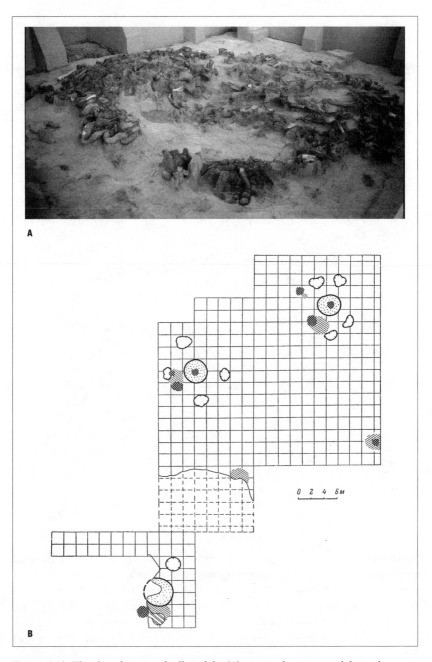

Figure 4.13 The foundations of village life: (A) ruins of a mammoth-bone house at Kostenki 11 (Russia) (photo by author); (B) floor plan of Dobranichevka (Ukraine), showing several former mammoth-bone structures (modified from I. G. Shovkoplyas, "Dobranichevskaya Stoyanka na Kievshchine," *Materialy i Issledovaniya po Arkheologii SSSR* 185 [1972]: 178, fig. 1).

Figure 4.14 The imposition of geometric form on the landscape is evident in the paved house floor at Le Cerisier (France), a nearly perfect square with entrances to the east and west. (After James R. Sackett, "The Neuvic Group: Upper Paleolithic Open-Air Sites in the Perigord," in *Upper Pleistocene Prehistory of Western Eurasia*, ed. Harold L. Dibble and Anna Montet-White [Philadelphia: University of Pennsylvania Museum, 1988], 69, fig. 3.5. Reproduced with permission from The University Museum, University of Pennsylvania)

France, sites along the Isle River contain pavements of stone that apparently reflect the shape of former structures. At Plateau Parrain, the pavement measures 16 × 18 feet in plan and seems to indicate an entrance on the south side. At Le Cerisier, the cobble flooring measures roughly 16 × 16 feet, creating a perfect square with opposing entrances on the east and west sides (figure 4.14).[103]

The house floors in the Isle River sites illustrate a pattern that becomes increasingly obvious and even somewhat oppressive in the millennia following the end of the Paleolithic—the imposition of geometric forms on the landscape.[104] Rectangles, squares, pyramids, and even the true, or Euclidian, circle are unknown in the natural world—biotic or abiotic.[105] Evolutionary biology yields symmetrical forms, but not geometric shapes (organisms tend to be lumpy and uneven). Simple geometric shapes are creations of the modern mind. From the late Upper Paleolithic onward, modern humans began to reconstruct the natural landscape in accordance with these mental creations, remaking the world in the form of the mind. So pervasive is the contemporary landscape of this externalized geometry that most of the global population now dwells largely within its own mind.

Geometry is a common theme in later Upper Paleolithic art. Although some of most impressive examples of representational art—both cave paintings and sculptures—date to after 20,000 years ago, abstract designs and geometric forms predominate in places like central and eastern Europe.[106] The meanings attached to the visual art of the later Upper Paleolithic remain unknowable, but the change in content and style from the EUP and the Gravettian is undeniable and suggests a shift in worldview. Later Upper Paleolithic people had acquired considerably more knowledge of and greater control over the external world than their predecessors, which surely was reflected in how they interpreted it. Perhaps they felt that they lived in a more orderly and manageable universe than had their ancestors. If so, they were indeed laying the foundation for the post-Paleolithic civilizations that would obsessively reconstruct that universe with numbers.

5

MINDSCAPES OF THE POSTGLACIAL EPOCH

[T]he [urban] revolution seems to mark,
not the dawn of a new era of accelerated advance,
but the culmination and arrest of
an earlier period of growth.

V. GORDON CHILDE

THE UPPER PALEOLITHIC was a time of more or less steady innovation and growth of knowledge. Although the uncertainties of discovery and dating in this range make it difficult to identify intervals when the pace of innovation may have slowed or halted altogether, the archaeological record yields a long succession of novelties from fur hoods to fishhooks and boats to baking ovens. The cumulative effect of these inventions was to redesign humans in a variety of ways, in accordance with the collective mind. Their practical impact from a biological standpoint was to allow one species—of recent tropical origin—to colonize most terrestrial habitats on the planet and to expand the size and density of some populations.

The development of the super-brain, which multiplied and spread with modern human foraging groups, also had brought forth a new phenomenon that operated above the level of the brain of an individual organism. The evolution of language and other forms of symbolic communication had allowed the members of each group to integrate their brains, significantly increasing the computational and recursive power of human thought. Moreover, by communicating thought from one generation to the next through oral tradition and artifacts, the collective mind had escaped the narrow confines of biological time and acquired a potentially unlimited "life span" of its own. Each collective mind had accumulated knowledge over the course of many Upper Paleolithic generations.

The growth in population and settlement size was a consequence of the redesigned human, who controlled an increasingly large and diverse supply of food and energy through technology. In the later Upper Paleolithic, modern humans in several parts of the world began to move away from their ancient mammalian foraging ecology to a sedentary economy based on domesticated plants and animals. The super-brain created an alternative ecological niche for the human organism—a niche subject to continual change and expansion that allowed for potentially unlimited growth. And with the emergence of large populations and sedentary communities, humans began to redesign the landscape as well as themselves.

During the Upper Paleolithic, efforts to refashion the landscape had been largely confined to the microhabitat represented by an artificial shelter or a house or by a corral or fish weir.[1] The domestication of dogs may be regarded as a modest and unintentional move to reshape a small piece of the biotic environment. But with the appearance of farming villages and the eventual spread of towns and cities, people began to transform meadows into crop rows and streams into linear canals. They constructed rectangular buildings, laid out orthogonal street grids, and designed geometrically complex plazas and palaces. These created landscapes had no precedent on Earth; they reflected neither the processes of geomorphology nor the evolved patterns of organic life, but the peculiar forms of the mind.[2]

As local populations grew to massive size—to thousands of individuals—people began to redesign their societies in equally strange ways. Forms generated in the collective mind were imposed on the evolved organic patterns of age, gender, and biological kinship. People came to be classified in accordance with created social and economic roles, such as stonecutter, potter, blacksmith, accountant, priest, slave, and king. This hierarchical specialization of thought and action apparently was essential to the functioning of the civilizations that arose from towns and cities in the Near East, China, Mexico, and other places in the postglacial epoch.

Although hierarchical social organization seems to have developed, at least in some places, without violence or armed coercion, warfare inevitably accompanied the early civilizations. More often than not, they faced real or perceived external threats that required a specialized hierarchy in the form of an army. Most early civilizations eventually expanded by forcibly incorporating cities and towns into a larger political entity. As a consequence, they also faced internal threats—uprisings, civil war, and disintegration into smaller units. Under these conditions, novel ideas and alternative realities often were categorized as dangerous and subversive, and the creative powers of the mind were suppressed.

The pace of technological innovation slowed in most early civilizations and sometimes ceased altogether, even though the total number of brains integrated into the collective mind of each state had increased exponentially. Moreover, many brains were devoted to specialized areas of thought and activity, and new means for their integration in both space and time had been created, especially writing. For reasons that remain obscure, western Europe broke out of the pattern of stasis after 1200, and the result was the modern world—in terms of both the advance of technology and the construction of reality.

The Civilized Mind

Ideology emerges with the state: a body of thought to complement a political entity.

BARRY J. KEMP

After more than seventy-five years of research and debate, the social and economic revolutions in human prehistory conceived by V. Gordon Childe remain key concepts in archaeology.[3] The first of them—the development of sedentary farming communities during the Neolithic—now seems less of a revolution and more of a gradual development from mobile foraging to controlled harvesting of plants and animals through cumulative technological creativity. The roots of the Neolithic extend back 20,000 years to the Last Glacial Maximum, if not earlier. The Urban Revolution—the birth of civilization—though, retains its explosive character.

Lewis Mumford labeled the early civilizations *mega-machines* and likened them to massive mechanical devices "composed solely of human parts."[4] They were, in many respects, experiments in social technology that reflected an inability to understand, or perhaps an unwillingness to acknowledge, the properties of the raw materials from which they were constructed. Many of the human parts of the machine probably were unhappy with their roles. Above all, civilization reflected the trauma of forcing nation-state organization onto a social order that was still defined primarily by family relationships.[5] Ideology was substituted for blood ties and marriage alliances. Inevitably, perhaps, the governments of the newly formed civilizations identified themselves with the forces that control the universe.

Civilization seems to represent a threshold in the density of the population or the size of the super-brain, at which the latter undergoes reorganization. There are, it would appear, simply too many individual brains and voices and insufficient communication and integration among them.[6] The size of the threshold is not entirely clear, but it numbers in the thousands at a minimum. Each of the cities that were hammered together to form a state in ancient China probably had several thousand inhabitants, and similar numbers are estimated for the urban centers that became the Maya civilization. But some of the latter, as well as some of the Sumerian cities that became *city-states* before being consolidated into a wider political organization, seem to have been larger.[7] And some sedentary societies that never morphed into nation-states, even though exhibiting elements of hierarchical organization, comprised many thousands of people.[8]

The reorganized super-brain became the *nation-state*. It exhibited a predictable hierarchical structure, with centralized political authority and a large organization of administrative assistants, priests, technicians, workers, and slaves, whose roles and relationships were at least partly defined by ideology. According to the Egyptologist Barry Kemp: "Fundamental to the state is an idealized image of itself, an ideology, a unique identity. It sets itself goals and pursues them by projecting irresistible images of power. These aide the mobilization of the resources and energies of the people, characteristically achieved by bureaucracy. We can speak of it as an organism because although made by man it takes on a life of its own."[9] The early civilizations also exhibited a pattern of specialization in thought and activity among their individual components. Even the simplest state organization contained craft specialists, such as potters, metallurgists, engineers, and scribes.

In order to function as an integrated whole, the national mind generated systems of writing, along with systems of weights and measures and of coinage. These were newly created internal components of the reorganized super-brain, analogous to spoken language and visual art in Upper Paleolithic times.[10] The earlier forms of communication and record keeping were apparently insufficient to integrate and manage such large organizations. Civilizations have continued to develop new means of management and integration, such as clocks, newspapers, radio, and the Internet.

Monumental architecture in the form of public buildings and plazas was another seemingly inevitable manifestation of civilization. Although typically regarded as the primary technical achievement of the early nation-states, the massive public structures also were a form of communication analogous to art and language in many ways. They were powerful symbols of the political and religious ideology of the state. And they were typically used as the setting for major public ceremonies or rituals.[11]

The social hierarchy of the early civilizations is so similar to that of the eusocial insects, especially the ants, that many of the same categories, such as worker and soldier, can be applied to both types of societies. The reorganized super-brain that represents the nation-state exhibits closer parallels to the super-organism than did its Upper Paleolithic and Neolithic predecessors. A fundamental difference remains, however, between the two entities. Ant societies build massive complexes of chambers and passageways, but they have no need for monumental architecture and public rituals. Such efforts would be a waste of time, energy, and resources. The reproductive biology of the eusocial insects has allowed them to generate state-like organization within the context of the family, so there is no conflict between king and clan.[12]

The slowing pace of innovation in the early civilizations—especially after what seems to have been an extraordinary surge of invention in the period leading up to, and perhaps immediately following, the formation of these societies—suggests that the newly reorganized super-brain was not very creative. The perceived lack of innovation in nation-states before the end of the Middle Ages in Europe has been a recurrent topic of debate among historians of technology.[13] After what Robert McC. Adams described as "a brief creative burst" and "extraordinary technological advances" associated with the rise of urban centers, inventions almost completely ceased in the ancient Near East.[14] The pattern is less evident during late antiquity in Hellenistic Greece and imperial Rome, and especially in China (before the Ming dynasty), but even in these settings there is a contrast with developments in western Europe after 1250.[15]

Both the organization and the ideology of the early nation-states probably discouraged novel thinking. Self-appointed governments are threatened by alternative realities, and the creative powers of individual brains would have been a potential source of instability in these hierarchical systems.[16] Moreover, the people at the top of these newly formed hierarchies had acquired the means to suppress alternative thinking—at least when externalized by speech or action—through the apparatus of the state. And they promoted the notion that they were connected to the controlling forces of the universe, either representing the divine or at least possessing special abilities (applied astronomy or other methods of divination) to interpret the direction of those forces correctly.

A major consequence of civilization was a slowing—and, in some cases, even an arresting—of the growth of the collective mind with respect to the accumulation of knowledge and the increased ability to manipulate the external world. The growth of the mind is dependent on creativity and novelty—applied to both technology and explanation. The narrow hierarchical structure of the early nation-states seems to have discouraged or suppressed much of it because the increased potential for creativity represented by a larger and more diverse population was more of a threat than an asset. The reorganized super-brain represented by the early civilizations was less capable of generating alternative realities than were its predecessors, and, for the most part, reality was stabilized for several thousand years.

An Alternative Landscape

[H]umans have been engaged in creating a living and working place, a
human-built world, ever since their ouster from the Garden of Eden.
THOMAS P. HUGHES

The Near East is the region where both sedentary farming communities and early civilizations first developed. The consensus among archaeologists is that the reason lies in its high biological productivity—enhanced by changing weather patterns at the end of the glacial epoch—and its mosaic of diverse local habitats, which supported a number of potential plant and animal domesticates. As Upper Paleolithic foraging groups expanded their control over this rich environment 20,000 to 15,000 years ago, their numbers increased with particular vigor.

The well-known Fertile Crescent—which stretches from the Jordan Valley north and east across Anatolia and then bends southeast toward the Persian Gulf (figure 5.1)—supported open woodland and grassland inhabited by various wild cereals (wheat, barley, and rye) and legumes (beans, peas, and lentils). Wild sheep and goats dwelt in the uplands. As today, there was significant altitudinal variation in local environments. After 17,000 years ago, temperature and moisture began to rise.[17]

The broad-based high-tech foraging economy that long preceded the farming villages of the Fertile Crescent is illustrated by an Upper Paleolithic settlement on the Sea of Galilee. Ohalo II (Israel) was occupied immediately after the peak cold and aridity of the Last Glacial Maximum, roughly 20,000 years ago. An unusual drop in water level two decades ago exposed traces of brush huts with stone flooring and other remains

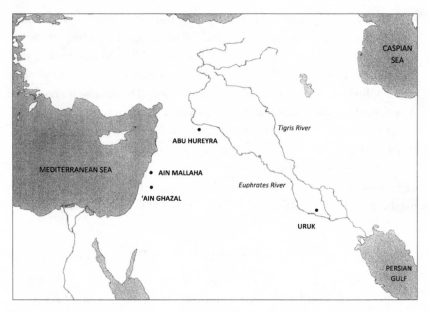

Figure 5.1 Map of the Fertile Crescent region, showing the location of sites mentioned in the text.

that had been exceptionally well preserved in the waterlogged sediment. Plant foods consumed at the site included more than thirty species, among which were wild emmer wheat and barley. Other plants apparently were used for medicinal purposes. Gazelles, birds, fish, and mollusks were represented among the animals. Technologies for preparing food included mortars, pestles, and an arrangement of stones that may have been a simple baking oven.[18]

People in the region continued to pursue a mobile foraging economy as the climate became warmer and wetter after 17,000 years ago. During this phase, groups expanded into areas of former desert. After 14,000 years ago, more stable settlements appeared. They include sites assigned to the *Natufian* culture, such as Ain Mallaha (Israel) and contemporaneous occupations at sites like Abu Hureyra (Syria). Many Natufian settlements are substantially larger than earlier sites and contain traces of circular semi-subterranean dwellings with stone foundations. The process of innovation continued, as Ofer Bar-Yosef observed: "The technological innovations introduced by the Natufians, such as sickles, picks, and improved tools for archery, were added to an already existing Upper Paleolithic inventory of utensils that included simple bows, corded fibers, and food processing tools such as mortars and pestles."[19] Although storage facilities are not especially common—Ain Mallaha contains some likely examples—burials are quite numerous in comparison with those of the preceding era, suggesting a more sedentary lifestyle. Also indicative of sedentism are household trash heaps containing the remains of mice.[20]

It appears that settled communities preceded, and in some respects triggered, agriculture, and not the reverse. Traces of domesticated plants do not show up in the Levant until after a sedentary economy was already in place. The cooler and drier climates of the Younger Dryas period (roughly 12,800 to 11,300 years ago) may have provided the catalyst. The expanding Natufian population probably experienced severe stress as the resource base contracted during this period. The cultivation of cereals like einkorn and rye seems to have been a response, and there is evidence for small-scale agriculture at sites on the margin of the retreating forest zone like Abu Hureyra. As conditions worsened, Abu Hureyra was abandoned altogether. It was reoccupied, however, at the end of the Younger Dryas by a larger community—a true farming village.[21]

Experimental research indicates that the process of domestication—altering the genetic structure of plants and animals through artificial selection—could have taken place in twenty to thirty years with some wild cereal species.[22] The intensification of harvesting and replanting as conditions deteriorated during the Younger Dryas may have accelerated the selection of desired phenotypes. Instead of adapting themselves to a changing environment, villagers at the western end of the Fertile Crescent were redesigning the environment—creating an alternative landscape. The consequences of domestication for human population biology were immense because the altered landscape could support more people per unit area. A larger population density would provide both the need and the means for further alterations; potential growth would be limited only by the total resources of the planet.

By 9,000 years ago, agricultural settlements like 'Ain Ghazal (Jordan) occupied more than 25 acres and contained multiroom and even multistory houses (figure 5.2). The dwellings exhibit a shift from the earlier circular form to the rectilinear shape that continues to dominate the mindscape. Grain-storage facilities and public structures—shrines and, eventually, temples—also were present in these sites. The population of 'Ain Ghazal is estimated at 2,500 to 3,000 people by 8,000 years ago. Both the domestication of sheep and goats and the production of pottery took place at this time.[23]

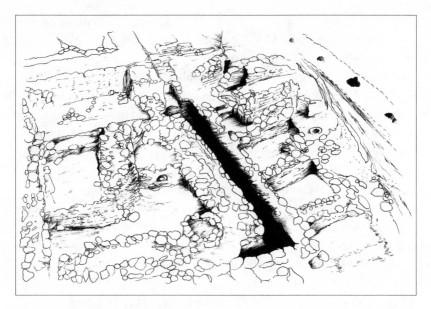

Figure 5.2 The settlement of 'Ain Ghazal (Jordan) occupied more than 25 acres by 9,000 years ago, comprising multiroom and multistory houses. (Drawn by Ian T. Hoffecker from a photograph by Y. Zo'bi)

At this point, if not before, farmers colonized the eastern terminus of the Fertile Crescent, establishing villages in southern Mesopotamia. Although drier than the Levant, the alluvial lowlands near the Persian Gulf are especially suited to irrigation due to the configuration of natural watercourses and the overall topographic setting. Between 7,800 and 5,800 years ago, the local population constructed and managed a growing network of irrigation canals. As the area of irrigated cropland expanded, the population and villages grew in size. Large-scale irrigation was another important step in redesigning the environment that had substantial consequences for the human population. It entailed significant alterations to the physical landscape, not simply to its plant and animal inhabitants, and expanded human knowledge in a new realm: hydrology and water engineering. Moreover, the larger canals eventually acquired another highly important function—facilitating transport and trade among the growing network of communities in the Mesopotamian alluvium (figure 5.3).[24]

The world's first cities emerged from the rapidly growing population centers of the Fertile Crescent. The rate of growth was exponential. The town of Uruk on the Euphrates River in modern Iraq occupied an estimated

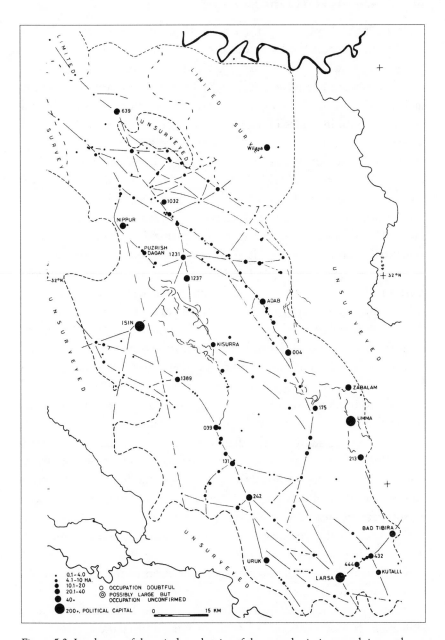

Figure 5.3 Landscape of the mind: as the size of the super-brain increased, internal connections—analogous to those of the neocortex in an individual brain—developed in the form of roadway and canal systems in southern Mesopotamia. Ur III-Isin-Larsa–period settlement patterns are shown here. (From Robert McC. Adams, *Heartland of Cities: Surveys of Ancient Settlement and Land Use on the Central Floodplain of the Euphrates* [Chicago: University of Chicago Press, 1981], 162, fig. 31)

175 acres by 5,800 years ago. It was more than five times the size that 'Ain Ghazal had been roughly 2,000 years earlier. Within 600 to 800 years, Uruk had grown to approximately 620 acres and contained a population estimated at between 25,000 and 50,000. By 4,900 to 4,800 years ago, the city had exploded in size to almost 1,480 acres.[25] The interactions within and among the growing cities of southern Mesopotamia—not only trade and alliance, but also competition and military conflict—became increasingly intense. Uruk and other cities were surrounded by defensive walls.

By 5,100 years ago (3100 B.C.E.), the cities had organized themselves into a social and economic hierarchy: the civilization of Sumer.[26] As during the earlier transition from a foraging to an agricultural economy, population stress induced by climate change may have been a catalyst. Drier climates prevailed after 5,800 years ago, and conditions became even cooler and drier around 5,200 to 5,100 years ago. A decline in food production and the abandonment of the most arid areas is widely thought to have created a mass of refugees and a potential pool of cheap dependent labor.[27] At the same time, the organizational demands of an increasingly complex society and economy seem to have spawned a class of administrators and record keepers, while armies demanded commanders and foot soldiers. And at the bottom of the hierarchy were a growing number of slaves, generated by military conquest.[28]

Major technological innovations accompanied the emergence of Sumerian civilization. The most notable were writing and mathematics, which seem to have evolved from the digital notational systems of the Upper Paleolithic. A significant breakthrough—analogous to Gutenberg's invention of movable type—was the development of numerical notation tablets around 5,400 to 5,300 years ago. Pictographic writing appears in southern Mesopotamian sites 5,200 to 5,100 years ago (figure 5.4).[29] Analysis of written texts reveals that the new information technology was used primarily for accounting and record keeping.

Writing and numerical systems seem to be essential to the nation-state, which is simply too large and complex to function without them, and therefore critical to the reorganization of the super-brain that coincided with their development. More broadly, writing and mathematics altered the collective mind in a fundamental way. The simple notational systems of the Upper Paleolithic represented an externalization of brain function on a very small scale. With the invention and widespread application of writing, a rapidly expanding corpus of memory was moved outside individual brains and into various technological forms of storage. Most of human memory now resides in these external components of the super-brain.

Figure 5.4 The hand was finally employed to create symbols in digital form, as does the vocal tract, with the invention of writing and mathematics, as illustrated by a Sumerian pictograph tablet from the settlement of Kish, dating to more than 5,000 years ago. (Drawn by Ian T. Hoffecker, from a photograph in Glyn Daniel, *The First Civilizations: The Archaeology of Their Origins* [New York: Crowell, 1968], 51, fig. 4)

By externalizing memory and creating a system of visual–linguistic feedback—analogous to the sensory feedback of artifact manufacture that began in the Lower Paleolithic—writing altered the process of thinking. The growing archive of memory and thought came to be accessible across increasingly large expanses of space and time. The super-brain had existed outside the lives of individuals in oral tradition and the corpus of information stored in artifacts (primarily in analogical form). Now it assumed a significantly greater presence as external representations were stored and manipulated in digital form. And as people began to discover the properties of numbers and apply them to calculations rather than simple counting, mathematics also changed the process of thinking.

The adoption of domesticated animals for agriculture and transportation was a major technological advance that preceded and probably accelerated the rise of cities and nation-states. Domesticated oxen were used to pull plows, ultimately derived from the Upper Paleolithic digging stick, for crop production, while domesticated donkeys were employed as pack animals. Both innovations were in place before 5,000 years ago.

They represented the first use of an energy source other than human muscle power. Donkeys are credited with intensifying interaction among the
Mesopotamian cities by facilitating trade and communication. Another
harnessed source of energy probably was wind power: there is evidence
for the presence of sail boats that—along with barges hauled by people
or oxen—were used to move large quantities of goods along the network
of waterways.[30]

The Sumerians traditionally are credited, of course, with the invention
of the wheel. The key innovation actually is the wheeled vehicle, which
is relatively complicated and was subject to many later refinements, such
as the revolving front axle.[31] Wheels were shaped from solid wood and
attached to heavy four-wheeled wagons, usually drawn by oxen, as well as
two-wheeled carts. Powered vehicles are tied to the expanding trade and
communication networks among cities and towns, but they also had military applications. Like the Upper Paleolithic bow, wheeled vehicles were
not modeled on or inspired by any observable natural phenomena. They
were an ingenious creation of the mind that yielded almost immediate and
highly practical benefits.[32]

Both raw materials and artifacts became commodities in Sumer, and
written texts document the production and transport of a variety of items.
This development was related to a fundamental change in the way people
made things, which reflected the hierarchical social and economic structure. Between 8,000 and 7,000 years ago, Mesopotamian pottery vessels
became increasingly standardized: "mass-produced" on revolving pottery
wheels.[33] Textiles also were woven on a titanic scale by 5,000 years ago
in Uruk and other major cities. Thousands of workers, typically dependent women and children, manufactured woolen textiles in state-owned
workshops reminiscent of a Charles Dickens novel. Cloth was woven on
horizontal and vertical looms and then treated with an alkaline solution
in large vats.[34]

For many archaeologists and historians, monumental architecture represents the most impressive technical achievement of the early civilizations. This view may be influenced by the psychological effect that these
structures still possess, despite their weathered familiarity and the long
absence of the formidable political entities they once symbolized.[35] The
shrines that had been constructed in south Mesopotamian villages were
expanded into large temple complexes—often associated with staged tow-
ers, or ziggurats—in Sumerian cities. A major temple in the city of Ur was
described by Samuel Noah Kramer:

[It] consisted of an enclosure measuring about 400 × 200 yards which contained the ziggurat as well as a large number of shrines, storehouses, magazines, courtyards, and dwelling places for the temple personnel. The ziggurat, the outstanding feature, was a rectangular tower whose base was some 200 feet in length and 150 feet in width; its original height was about 70 feet. The whole was a solid mass of brickwork with a cover of crude mud bricks and an outer layer of burnt bricks set in bitumen. It rose in three irregular stages and was approached by three stairways consisting of a hundred steps each.[36]

The temple complexes were the locus of regularly scheduled public ceremonies.

Although the Sumerian temples and ziggurats were the first known monumental buildings, it was the Old Kingdom of Egypt that designed and built the most awesome structures among the early civilizations—in terms of both scale and precision. The Great Pyramid at Giza, which was constructed during the reign of Cheops in the Fourth Dynasty (2600–2450 B.C.E.), is the supreme example. Assembled from roughly 2.3 million stone blocks (averaging 2.5 tons apiece), the base rests on a nearly perfectly level platform that occupies 13.1 acres. Each side originally measured between 755.43 and 756.08 feet in length (that is, varying by no more than 7.9 inches, or 0.01 percent). The four corners were almost perfect right angles (less than 1° deviation), and the entire structure was aligned almost exactly with true north.[37]

The Great Pyramid and other monumental buildings of the ancient civilizations were an early example of the application of mathematics to technology. The Sumerian irrigation systems—reportedly laid out with the help of levels and measuring rods—were another. The applications extended beyond simple arithmetic to geometry. It was a profound development with far-reaching consequences—virtually all of modern industrial technology is produced with mathematical applications—and a defining difference with earlier technology. In part, it reflected practical requirements imposed by the scale of monumental structures and other public works such as roads and canals. The parallel applications of writing and arithmetic to record keeping and accounting addressed analogous requirements of scale in the national economy. They also reflected a more precise rendering of the mindscape.

Mathematics was applied to time as well as space, and it represented a new way of perceiving and interpreting the universe. All the early civilizations designed long-term calendars, which organized time hierarchically into numerical units. The Sumerian calendar subdivided the year into twelve lunar months and required the periodic insertion of an intercalary month to adjust for the imprecise synchronization of the solar and lunar cycles. The day was subdivided into hours, which could be measured with a water clock. The Egyptians developed several unsynchronized calendars that covered annual cycles of flooding of the Nile River, as well as lunar and stellar cycles, but they also devised a civil calendar of twelve 30-day months and 5 extra days, for a total of 365 days. The day was subdivided into twenty-four hours, but the length of the hours varied according to season.[38]

Agriculture and nation-states emerged in several other parts of the world after the appearance of Sumerian civilization 5,100 years ago. Each civilization was unique and reflected both local environmental conditions and historical factors, but all of them exhibited a general pattern of hierarchical social organization and economic specialization—as well as new information technologies for the integration and management of its components—that may be explained as a common response to very large densities of people.

Egyptian civilization materialized shortly after the appearance of the Sumerian state, but as the result of a somewhat different process. The autonomous city-states of southern Mesopotamia never developed in the Nile Valley, and political integration was achieved through military unification of the northern and southern regions: Lower and Upper Egypt, respectively. The notion of the kingdom as "two lands" remained ever after. Egyptian civilization nevertheless produced most of the features found in Sumer.[39]

A nation-state did not arise in China until roughly 4,000 years ago: a full millennium after the appearance of Sumer. The roots of sedentary settlement and farming in East Asia are deep, however, extending back to the making of pottery and harvesting of potential plant domesticates in the aftermath of the Last Glacial Maximum. Settled communities based partly on agriculture and similar to those in the Near East about 9,000 years ago were present in northern China by 8,000 years ago. During the next few millennia, the population continued to grow, and by 4,000 years ago northern China was filled with walled towns, each inhabited by several thousand people under the control of a local clan. Many of them were brought under the aegis of the first ruling dynasty (Xia) to create a nation-state. It exhibited most of the features of Sumerian and Egyptian civilization—with the notable exception of monumental architecture, which was present but in subdued form.[40]

Civilization arose much later in the Americas. One reason for the delay may have been the absence of a lengthy development of a broad-based high-tech economy during the later Upper Paleolithic that produced semi-sedentary settlements before the beginning of the postglacial epoch. Such a pattern of development is now evident in both the Near East (including the Nile Valley) and China. Modern humans do not seem to have reached the Western Hemisphere until relatively late (about 15,000 years ago), and the American contemporaries of the Natufians appear to have been mobile foragers with a relatively simple technology. Farming villages were present in parts of Central America by 3,800 years ago, and complex societies with elaborate tombs and monumental art emerged a few centuries later.[41]

Civilization on a scale commensurate with that in Egypt or China did not appear in Mesoamerica until 2,300 years ago with the late Preclassic Maya, who developed writing, monumental architecture, and a calendar. The Maya comprised a large and dense population; the city of Tik'al is thought to have been inhabited by as many as 90,000 people in the Late Classic period.[42] Tik'al and other cities contained massive temple complexes and pyramids that are remarkably similar to those of the early civilizations in the Near East, although they undoubtedly reflect independent invention. By 600, civilization also had emerged in South America. The Andean states are distinguished from others by the absence of writing, although accounting and record keeping were performed with a system of knotted cords (quipu).[43]

All the complex societies that developed in the postglacial epoch redesigned their environmental settings in various ways. This began with the reorganization of plant and animal communities into gardens, crop fields, herds, and livestock pens. In many places, people created artificial waterways (canals), springs (wells), and ponds (reservoirs). But the most bizarre manifestations of this pattern were the *created social environments* of urban centers. Beginning with villages and towns, people built rectilinear houses with rectangular windows and doorways. The cities often contained orthogonal avenues and causeways, as well as large public buildings and plazas that were pure creations of the collective mind—geometric forms that bore little resemblance to features in a natural landscape. In some places, their immense size and the time and labor invested in their construction, reached the level of the absurd. They seem to have served no useful function from the perspective of human biology; the public ceremonies held at the temple complexes at Ur, Giza, or Tik'al could have been performed in an open field. Rather, they were symbols of the equally immense political structures that now dwelt on the Earth.

An Age of Stone

[A]ll knowledge of mind is historical.
R. G. COLLINGWOOD

R. G. Collingwood observed that the early civilizations had no concept of history. The "habit of thinking historically," he argued, was a surprisingly recent phenomenon, dating no earlier than the eighteenth century.[44] The first nation-states instead saw themselves through the prism of myth, or what Collingwood labeled "theocratic history." In both cases, past events were explained by the actions of supernatural beings.

The absence of a historical perspective in the early civilizations seems to reflect a cultural construction of time that was cyclical, rather than linear. It was a perspective both common and logical; time was perceived as a phenomenon that revolved continuously, as do the days, lunar months, and seasons. Such a perspective does not preclude history, but it does not demand or even imply it in the same way as does a linear concept of time. The modern linear construction of time, which has its own theological basis, developed in late antiquity.[45] Unlike a historical narrative composed of an irreversible sequence of past events, a myth—as Claude Lévi-Strauss wrote—"explains the present and the past as well as the future"[46] and is, as he famously suggested, an instrument "for the obliteration of time."[47]

A preference for myth over history also reflects an aversion to change and would seem to be another manifestation of the reactionary and repressive character of the early civilizations—or, at least, their consistent failure to cultivate new ideas. By doing so, they slowed and, at times, almost stopped the process of history, especially if we define it in Collingwood's words as "the history of thought." The few available fragments of historical narrative provided by the early nation-states are largely confined to lists of kings and accounts of battles or natural catastrophes. There was no sense of progress toward a better world—a view that pervades modern life.[48]

Among the early civilizations, Egypt was the most successful in suppressing change and obliterating time. Despite major disruptions—which included the collapse of the Old Kingdom, foreign invasion and occupation, and a deeply disturbing challenge to religious doctrines (Akhenaten's monotheistic "counter-religion" in 1352 B.C.E.)—the culture of ancient

Egypt endured with minimal alteration for almost 3,000 years. The Egyptians steadfastly protected their traditional worldview—and the forms of language, art, and ritual through which it was expressed—from internal subversion and foreign contamination. As Jan Assmann observed, an educated Egyptian of the Roman period could have read and understood a Third Dynasty tomb inscription composed more than two and half millennia earlier. The longevity of Egyptian culture is without parallel and can be regarded not as a deficiency, but "rather as a special cultural achievement—indeed, one perhaps unique."[49]

The familiar pattern of Egyptian ideology and art arose during the Early Dynastic Period and the Old Kingdom (3100–2150 B.C.E.). The construction in the early Third Dynasty of Djoser's step pyramid (figure 5.5), which monumentalized the politically important *sed* festival, seems to have marked a critical turning point.[50] But the earlier creation of hieroglyphic writing, which allowed the preservation of sacred texts, also was significant. The Egyptians insisted on repetition of the precise wording of these texts in ritual settings. This reflected a belief that the proper functioning of the universe could be ensured only by careful adherence to the form of the sacred.[51]

Ancient Egypt was a paradox. It became the most extreme example of conservatism among nation-states, but only after having become one of most revolutionary societies in history. More than any other early civilization, post-unification Egypt seems to have attempted a rapid replacement of the traditional organization of clans and villages with the idea of the nation. The government of the Old Kingdom implemented a social

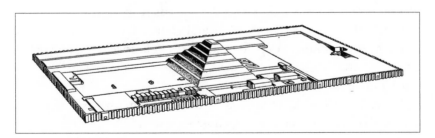

Figure 5.5 The construction of Djoser's step pyramid within an enclosure during the Third Dynasty in Egypt linked the *sed* festival, which celebrated the continued rule of a pharaoh, to monumental architecture and seems to have played a role in the emergence of a nation-state ideology. (From I. E. S. Edwards, *The Pyramids of Egypt* [Harmondsworth: Penguin, 1961], 54, fig. 5. Reprinted with the permission of Penguin Books)

revolution from above that suggests parallels with the Soviet Union of the 1930s or other modern states that have forced radical social changes on the population. Egypt was divided into provinces (*nomes*) controlled by administrators that bore no relation to the earlier pattern of competing chiefdoms.[52] Perhaps this is why the ideology and public symbolism of the Old Kingdom seems excessive.

The revolution faltered at the end of the Sixth Dynasty (2150 B.C.E.), when the Old Kingdom collapsed and Egypt reverted to a group of regional political centers. Centralized government was restored in 2040 B.C.E., but unrest and civil war erupted within a few decades and lasting stability was not achieved until 1991 B.C.E. under the Twelfth Dynasty. The Middle Kingdom resuscitated the religio-political doctrines of the Old Kingdom and publicized them in new ways that paralleled the governmental propaganda of modern nation-states. The preceding period was officially recalled as a nightmare of chaos. A "police state" atmosphere prevailed.[53] The pattern again is reminiscent of the Stalin era: the reactionary conservatism of a successful revolutionary movement.

By contrast, Sumer had developed in a very different way from Egypt, with the emergence of autonomous city-states. The social and economic hierarchy that represented early Sumerian civilization had arisen largely, it seems, through the "labor revolution" described earlier, including the managerial demands that accompanied it, and was not imposed from above.[54] Perhaps the autonomous urban centers would have maintained the steady pace of innovation that preceded and accompanied their formation, but within a few centuries they were forcibly incorporated into a larger nation-state—first by one of the south Mesopotamian kings (2375 B.C.E.) and several decades later by Sargon, who united northern and southern Mesopotamia under the Akkadian state. The region was subsequently controlled by others, including the Babylonian and Persian empires. Although there were some advances in knowledge, especially in astronomy and mathematics, the rate of progress seems slow in comparison with that in the earlier period.[55]

The early civilizations of Eurasia were disrupted in various ways by the development and spread of iron technology after 1200 B.C.E. The critical innovations emerged beyond their borders and control. Once the technical challenges of iron smelting were overcome, the widespread distribution and accessibility of iron-ore deposits ensured the mass-production of cheap and effective weapons and farming tools. Another innovation from the uncivilized fringe that had both military and economic consequences was the domestication of the horse. Mounted cavalrymen with iron weapons toppled pharaohs and administrators, as states and empires rose and fell during the many centuries of conflict that followed. Iron technology also transformed Africa. Bantu-speaking peoples with iron implements had occupied most of the lands south of the Sahara by 1000 C.E.[56]

The later civilizations of Eurasia were less effective at suppressing change and innovation—especially in comparison with ancient Egypt. Among them, Greece made the most impressive contribution to the growth of the mind, and it may be significant that it began as a collection of autonomous city-states around 600 B.C.E. Inevitably, perhaps, they were incorporated into a larger political entity, initially as a result of the Peloponnesian War, which ended in 404 B.C.E., and subsequently by Philip of Macedon and his son Alexander, whose death in 323 B.C.E. is equated with the beginning of the Hellenistic period. And although the principal contributions to knowledge are usually attributed to the city-state phase, the most significant technical achievements date to the Hellenistic period.

During the earlier phase, there were some advances in architecture, which included the application of knowledge about optics and acoustics, and innovations in mechanical military equipment such as catapults. But the Hellenistic period (323–146 B.C.E.) produced a series of major breakthroughs in mechanical engineering that laid the foundations of industrial technology.[57] Among them were levers, screws, gears, springs, valves, hydraulic pumps, and compressed-air devices. These technics were accompanied by brilliant mathematical work, which—anticipating Galileo—included the application of numbers to mechanics by Archimedes (287–212 B.C.E.).[58]

Extenuating circumstances may have created a favorable environment for novelty and innovation in Greece, even after the autonomous city-states were subsumed under a central government. Much of the creative thinking of the Hellenistic period took place in Alexandria, a large and ethnically diverse city ruled for eighty years by the first three kings of the Ptolemaic dynasty, who actively promoted the growth of knowledge. The massive library at Alexandria represented another major contribution to the latter. It was an accessible storehouse of external memory on an unprecedented scale. Innovations also came from other cities that, like Alexandria, were outliers or colonies.[59]

But the application and dissemination of the engineering innovations of Archimedes and others were limited, and they did not have a major impact on Hellenistic society and economy. The primary applications seem to have been confined to ship design and military machinery. One constraint may have been a commonly expressed preference for abstract theory over practical use. Ultimately, these discoveries had little effect on how the ancient Greeks interpreted the world; they never developed the mechanistic outlook of the European clock makers.[60]

The history of Chinese civilization probably yields the most insight into the relationship between the organization of the nation-state and the growth of knowledge. This is because both the pace of innovation and the degree of political centralization varied significantly during the course of Chinese history, which followed a different path from that of western Eurasia. There were few or no major inventions in the millennia leading up to the formation of a nation-state around 2000 B.C.E.,[61] and Chinese civilization began with something of a technological deficit relative to Sumer. But in the centuries that followed, key innovations were broadly applied, with profound effects on society and economy. The most significant were the development and spread of iron agricultural implements, including the plow, and irrigation systems, which took place during the Eastern Zhou period (770–221 B.C.E.) with substantial impact on agricultural productivity.[62]

The national political structure underwent dramatic changes before and during the Eastern Zhou period. The first two dynasties, Xia and Shang, had consolidated many but not all of the numerous towns scattered across northern China. The social hierarchy of these towns was based primarily on family relationships. The Eastern Zhou state disintegrated into smaller competing entities during the Warring States period (450–221 B.C.E.), recalling the collapse of the Old Kingdom in Egypt or the city-states period in ancient Greece. Like the latter, it is associated with a burst of creativity in philosophy and literature, as well as with technological advances.

China was reunified in 221 B.C.E. under the Qin dynasty, which embarked on a major expansion of central government power—crushing local landlords, establishing a national system of weights and measures, and even inaugurating monumental construction projects, including the Great Wall. The emperor promulgated an explicit ideology, Legalism, and burned books and executed scholars who expressed alternative ideas. But the Qin dynasty was short-lived, and China resumed its accumulation and expansion of knowledge under a succession of dynasties that exerted varying degrees of central control over the population.[63]

After 700, the rate of growth accelerated in many spheres of technology and China began to move toward an industrial economy far in advance of any other civilization in the world. Earlier developments in iron production led to coal-fired blast furnaces in northeastern China. In 1078, tax officials recorded a total production figure of 125,000 tons of smelted iron (figure 5.6).[64] The earlier irrigation systems had led to advances in hydraulic engineering and water-powered machinery, such as trip hammers and bellows for the smelting furnaces. Moving water also was used to power Su Sung's famous mechanical clock of 1092. There were innovations in textile manufacturing, including the mechanical cotton gin and water-powered spinning frame. All these developments were accompanied by a dynamic commercial economy and an expanding overseas trade, which was facilitated by the most advanced ships in the world, constructed with watertight bulkheads and navigated with the aid of a magnetic compass.[65]

The incipient industrial era came to end about 1400. The innovations largely ceased, and the activities of the merchant fleet were curtailed. China

Figure 5.6 China appeared to be in the early stages of an industrial age roughly 1,000 years ago—and far in advance of Europe—with accelerating iron production, intensifying use of water-power technologies, and international overseas trade. This depiction of a smelting furnace with water-powered bellows dates to about 1300. (From Joseph Needham, *Science and Civilization in China*, vol. 4, part 2, *Mechanical Engineering* [Cambridge: Cambridge University Press, 1965], 371, fig. 602. Reprinted with the permission of Cambridge University Press)

fell behind other civilizations technologically and eventually became vulnerable to smaller nation-states like Britain and Japan.[66] The suppression of cultural change at this time is widely attributed to the policies of the Ming dynasty, which took over in 1368, used central government bureaucracy to discourage the growing power of industrialists and merchants. Foreign trade was prohibited in 1371 by imperial decree, and the construction of large ships was banned in 1436. Similar policies were continued by the succeeding (and final) Qing, or Manchu, dynasty.[67]

The timing of the government crackdown may have been critical. Despite the widespread effects of the accelerating innovations in technology and their applications, the emerging pattern of industrial production and expanding commerce did not have a lasting impact on Chinese society and thought. Had continued growth been permitted, or even encouraged, by the central government, perhaps the modern world would have been largely a creation of the Chinese mind.

The Modern World as a Change of Mind

[T]here is an immediate and close connection between Christian ideas of creation and of the Kingdom of God and the development of modern technology. There is also a connection between Christian hope and the technological utopia of the future.

ERNST BENZ

The central problem in the history of the modern world is why and how western Europe broke out of the creative stasis into which all civilizations fall sooner or later. The explanation lies in understanding how the creative powers of the mind, specifically of individual brains, can operate freely within the hierarchical structure of a nation-state. There is no obvious or simple answer to the problem, but—like the earlier emergence of farming settlements and civilizations—both environmental variables and specific historical factors seem to have played a role in what happened in Europe after 1200.

Western Europe occupies a unique position in the Northern Hemisphere with respect to temperature and precipitation. Clockwise currents in the North Atlantic Ocean ensure mild and moist climates despite comparatively high latitude. As a result, biotic productivity is higher in this region than anywhere else in northern Eurasia. In 1200, agricultural productivity and population density were rising due to the effects of a warm-climate oscillation (Medieval Warm Period). The available food surplus also seems to have increased, and a higher proportion of the population could pursue activities other than farming.[68]

In addition, the political and cultural landscape was unusual and possibly unique. Western Europe was divided into a chaotic tapestry of kingdoms and smaller political entities that had once been incorporated into the mighty Roman Empire. No central authority existed, and—as some historians have observed—no individual ruler or government had the power to suppress threatening novelties throughout the region, as did the Ming dynasty.[69] At the same time, shared affiliation with the former Roman Empire provided a common language (at least for the literate) and church organization. The components of a large and diverse super-brain were integrated.

Several historians, most notably Ernst Benz and Lynn White, argued that the origins of the industrial civilization of Europe lay in the theology of Western Christendom. They emphasized the peculiar attitude toward the natural world articulated by many in the Western church, who, rather than seeking to live in harmony with it, viewed nature as a potential domesticate to be exploited, controlled, and, if possible, redesigned for human benefit.[70] And far from being disdained, manual labor and practical achievements were regarded as virtuous. A deeply rooted Judeo-Christian construction of time as linear and progressive, rather than cyclical, may have created a more favorable climate for novelty.[71]

In any case, after centuries of stasis in farming techniques and technology under the Roman Empire, a succession of innovations after 700 laid the foundation for an "agricultural revolution" in Europe. They included improvements in the plow, toward a heavy wheeled design with a moldboard for slicing through turf, and the shift from oxen to more efficient horses for drawing it. And the use of horses required the development of a new type of collar and iron shoes to protect their hoofs from damage, especially in moist climates. Another critical innovation was move from Roman two-field crop rotation to a three-field system, which reduced fallow land from 50 to 33 percent each year.[72] The combined effect of these changes seems to have been substantial increases in crop yields during the centuries leading up to and after the beginning of the Medieval Warm Period, which saw further increases as climates improved after 1000.

Historians have long debated the role of slavery in inhibiting technological innovation and, more broadly, social and cultural change among the early civilizations down through late antiquity in western Eurasia (that is, the Roman Empire). As slavery declined, the argument goes, incentives for new labor-saving technologies and general improvements in economic efficiency would have risen. Lewis Mumford thought that the critical issue was replacing human muscle with alternative sources of energy.[73] In western Europe, such a trend became apparent by 800, with the growing application of water and wind power.

The technology for hydropower had been developed much earlier than the eighth century. There is evidence of water wheels in ancient Greece as well as in the Roman Empire. Indeed, the Romans made some improvements in their design.[74] The difference lies in the widespread use of water power during the later Middle Ages. By 1086, tax records indicate no fewer than 5,624 water wheels spinning in southern England alone. The generated mechanical energy was employed primarily to grind grain, but also saw other applications, including to power trip-hammers. In some coastal areas, where stream gradients were low, tidal mills were constructed.[75]

The windmill, though—first reported in Iran by 700—was a post-Roman invention.[76] Although wind power had been used to propel boats and ships for several thousand years, the idea of applying it to the generation of mechanical energy was apparently derived from knowledge of water wheels. The effective adaptation of the technology to moving air required the development of a revolving pedestal in order to adjust the position of the wheel to shifting wind directions. In western Europe, windmills were widespread before 1200 (figure 5.7). No fewer than 120 of them were operating near the Flemish town of Ypres by the end of the thirteenth century.[77]

The mechanical technology that Mumford termed the "key machine" of the industrial age—the weight-driven clock—was engineered somewhere in western Europe around 1250.[78] Its advent and far-reaching consequences for technological development and worldview were mentioned in chapter 1. In one respect, the weight-driven clock was regressive because it ultimately relied on human muscle power to provide the energy. The escapement mechanism, designed to release the stored energy, was a major technical achievement, however. And it was the first machine built entirely of metal. The chief incentive apparently was to create a device for the timing of prayers in monastic orders that was more accurate and reliable than the water clock.[79]

The impact of the mechanical clock on the human imagination was considerable. Clocks were, as David Landes remarked, "like computers today, the technological sensation of their time."[80] The clock makers became a new category of technical specialist, and their knowledge and skills found applications to other technologies, such as water wheels. Clocks eventu-

rois faulli en pies fa chambze eft uerfee
l a pzife fa lance la char ı a boutee
oʒs del enging la mife ʒ gtre mont leuee
ı oıfiel fameıllous ont la char efgartee
l tentent gtre mont facoullent la uolee

Figure 5.7 Both water- and wind-powered technology became increasingly widespread in western Europe during the late Middle Ages. (From Joel Mokyr, *The Lever of Riches: Technological Creativity and Economic Progress* [New York: Oxford University Press, 1990], 45, fig. 10. Reprinted with the permission of Oxford University Press)

ally spread to every community and created an arbitrary temporal matrix within which virtually everyone functioned. They generated the temporal component of the daily landscape of the mind. They became a model and metaphor for the entire universe.

The work of the clock makers was enhanced by another important innovation before 1300: eyeglasses.[81] The invention of spectacles reflected expanding knowledge of the properties of glass and how to manipulate them to achieve yet another step toward remaking the human organism. Yet despite the discernible momentum in harnessing new sources of energy and inventing new technologies, western Europe experienced a slowdown and something of a backlash during the fourteenth century. Catastrophic death and population decline were caused by severe famine in the second decade of the century, followed by outbreaks of plague, especially the so-called Black Death (1348–1350), which wiped out at least one-third of the population of Europe. Wars and violent uprisings erupted. There was a surge of interest in religious mysticism and a corresponding obsession with witchcraft.[82]

The pace of innovation resumed during the fifteenth century, which became the critical turning point for the emerging construction of reality that became the modern world. The invention of movable type and the printing press in 1455 had consequences on the same scale as those of the mechanical clock. Within fifty years, more books were produced than had been in the preceding thousand years.[83] Information was communicated among the disparate elements of the collective mind, stimulating new thought, and stored in readily accessible form for future generations. It was the most significant development in information technology or neuro-technology since the creation of writing.

Another innovation with fateful consequences was the adoption of gun-powder, probably from China, and its application to the propulsion of projectiles. Despite many advances in the catapult by the ancient Greeks and later improvements in the crossbow, the fundamental design of projec-tile weaponry in 1200 had not changed since the late Upper Paleolithic. A heavy projectile propelled by an explosive force in a metal tube represented a potentially major increase in power and range. The earliest cannons are reported from the early fourteenth century; during the fifteenth, they became larger, more mobile, and more common. Competition and con-flict among the fragmented political entities of western Europe ensured a perpetual arms race that yielded regular improvements in military technol-ogy.[84] And armies in western Europe were deployed in marching forma-tions with increasing precision, reproducing the hierarchical organization of the military in simple geometric shapes on the landscape.

European artillery and small arms also ensured an overwhelming mili-tary advantage over other peoples. Firearms played a role in the explora-tion of the world (in conjunction with diseases), which was initiated in the fifteenth century. Although the impact on non-European local popula-tions and cultures was catastrophic, and the economic benefits were prob-lematic, the people of western Europe became the first to acquire a knowl-edge of the planet as a whole. Like the mechanical clock, the expanded concept of the world had a powerful effect on their imagination. Some improvements in ship design and navigation were essential, such as the axial rudder and compass (also from China), but Fernand Braudel argued that the most important factor was psychological: overcoming the fear of sailing across vast expanses of ocean like the Atlantic.[85] This reflected per-haps a wider change in attitude toward the natural world.

The relationship between the acquisition of knowledge through both technological innovation and contact with other peoples and the construction of reality is illustrated on a grand scale in Europe between the fifteenth and seventeenth centuries. The accelerating pace of innovation after 1200 not only yielded a new body of knowledge about materials and processes of the surrounding world, but tipped the balance toward a novel construction of reality based on the model of the machine. By the end of the seventeenth century, Lewis Mumford wrote, "there existed a fully articulated philosophy of the universe, on purely mechanical lines, which served as a starting point for all the physical sciences and for further technical improvements."[86] The "scientific revolution" was led by a small group of people scattered across western Europe, including Georg Bauer (Agricola, 1494-1555) in Germany, Galileo (1564-1642) in Italy, Descartes (1596-1650) in France, and Francis Bacon (1561-1626) in England. For the first time since the Lower Paleolithic, the function of the sensory hand was understood and the acquisition of knowledge was tied explicitly to physical experimentation: technology applied to science. Whenever possible, processes were described mathematically; when necessary, mathematical knowledge was expanded for this purpose, such as the invention of calculus.[87]

To some extent, the Europeans redesigned the mind itself. Although brain function had been altered technologically since the creation of mnemonic devices in the early Upper Paleolithic, the creative powers of the mind either had been left free to roam, although always constrained by social setting, or had been suppressed by priests and police. The revolutionary scientists recognized the need to manipulate creativity by testing alternative realities under controlled conditions and by minimizing the subjectivity of the brain. The early civilizations had applied mathematics to technology—now it was applied to thought.

The rewards of creativity finally were recognized, even by kings, and a previously unknown sense of forward momentum was emerging. In 1306, Dominican Fra Giordano of Pisa noted both the benefits and the novelty of spectacles, invented within his lifetime, and remarked that "not all the arts have been found. . . . Every day one could discover a new art."[88] The sense of progress was a new form of super-brain consciousness that reflected an awareness of historical or cultural rather than biological linear time. Not only could the enlightened and innovative present be distinguished from the ignorant and backward past, but an alternative future reality—a paradise of knowledge—could be imagined with confidence. Francis Bacon was particularly outspoken in his vision of technological progress. He saw not only the limitations of the past, but also the potential of the future, describing a world with submarines, aircraft, research labs, telephones, and even air-conditioners.[89]

People Without History?

[S]ome savages have recently improved a little in some of their simpler arts.

CHARLES DARWIN

As European explorers spread across the oceans and disembarked on strange shores, they encountered a wide range of people previously unknown to them. In Mexico, they confronted a technologically primitive civilization with a major urban center. In the central Pacific, they saw chiefdoms composed of thousands of individuals. In the Arctic, they were impressed with people who inhabited large villages without benefit of agriculture. In some places, they discovered nomadic foraging societies equipped with very simple technology.[90]

Europeans of the eighteenth century were especially struck by the native people of Australia, who had little in the way of clothing or equipment. They came to be viewed as the most primitive of all peoples. When William J. Sollas published *Ancient Hunters and Their Modern Representatives* in 1911, he presented them as typical of Middle Paleolithic (and non-modern human) foragers. Even more primitive, Sollas asserted—and evocative of the Lower Paleolithic—were the already extinct natives of Tasmania (a large island off the southeastern coast of Australia).[91]

The discovery of a wide range of societies and cultures—from the very small and simple to the large and complex—had a powerful effect on the European imagination. Simple foraging societies like the native Australians aroused special fascination; they seemed to be "living fossils," offering a glimpse of earliest humanity. They provided further support to the view of history as progress. Indeed, it would appear that encounters with "primitive societies" had as much, if not more, impact on thinking about social and cultural evolution as did the discovery of ancient artifacts and the recognition by the early nineteenth century that they, too, could be organized into a sequence of evolutionary progress.[92]

The existence of such societies posed a question in this context: Why had some cultures failed to develop agriculture, power technology, and large settlements? Why had some people remained at what seemed to be a Middle or even a Lower Paleolithic level of society and culture? A widely accepted answer to this question in the nineteenth century was that the contrasts in technology and organization between modern and premodern societies reflected biological differences: primitive societies were the product of inferior races.[93] Such a view not only was morally objectionable, but was then and remains without any credible supporting evidence. In the years since their initial encounters with European explorers, most of these populations have adopted contemporary technologies and become part of complex societies. The global differentiation that began with the movement of modern humans out of Africa 50,000 years ago and their dispersal throughout the world has been in reverse for several centuries. As this trend continues, the differences among various peoples that seemed so fundamental to the Europeans will be understood as the consequence of historical factors.

An alternative explanation was *environment*: foraging peoples and other simpler societies encountered by European explorers inhabited places with severe limitations on growth and development. In many cases, this seems to be a valid explanation. Recent hunter-gatherers often have been found in desert margins, arctic tundra, or other environments where the potential for agriculture and population growth is constrained by aridity or cold. There are exceptions to the pattern, however. Some foraging societies of the recent past have occupied habitats with clear potential for domestication and sedentary settlement. And anecdotal evidence suggests that some of these foragers were well aware of the possibilities and requisite techniques for the cultivation of plants. They simply preferred their traditional way of life.[94]

One reason why Europeans misjudged foraging societies during the seventeenth and later centuries is that they lacked a historical perspective on them. With a better knowledge of the archaeological record of modern humans in Australia, southern Africa, northeastern Asia, and other regions, it becomes apparent that all of them have changed over time. With the advent of the mind and the spread of modern human groups, truly static cultures like those of the Lower and Middle Paleolithic vanished. The native Australians, for example, who comprise a diverse array of people living in a variety of habitats, have experienced nearly continual change since their arrival more than 40,000 years ago.[95] Even groups equipped with the simplest technology have created novel implements that probably were unknown in the early or middle Upper Paleolithic.[96]

At the same time, rates of change clearly have varied considerably. A slow pace of innovation and change may indeed reflect local environmental variables. But historical factors also seem to be at work. Roughly 2,000 years ago, for example, people in the Bering Strait region engineered some complex technology related to whale hunting that had significant social and demographic consequences. Their neighbors farther east—known to archaeologists as Late Dorset—did not , and there is evidence that they had actually discarded some of their earlier innovations, such as the bow and arrow.[97] Like nineteenth-century China, the Late Dorset people fell prey to more technologically advanced foreigners, at least partly as a result of their own historical trajectory. Even without a rigid hierarchy and centralized government to suppress innovation, the Late Dorset seem to have been particularly resistant to change; perhaps this also was a significant factor among other foraging societies.

6

THE VISION ANIMAL

The people themselves are friendly and intelligent,
with a good sense of humor. Though fond of relaxation,
they're capable of hard physical work when necessary.
Otherwise, they don't much care for it—but they
never get tired of using their brains.

THOMAS MORE

AT THE TIME THAT ANATOMICALLY modern humans evolved in Africa, the social insects were arguably the dominant form of terrestrial animal life on Earth. The ants alone are estimated to have achieved a biomass roughly equal to that of the current human population (more than 6.5 billion people). The social insects had assumed control of the most favorable nesting sites, forcing solitary insects into marginal zones. Their success is attributed to their organizational adaptations, particularly impressive among the eusocial, or "true social," insects such as bees and ants. The ants are represented by thousands of species and a wide variety of social adaptations.[1]

The eusocial insects exhibit many parallels with modern humans. The honeybees are noted for their rare ability to externalize representations, which they do with body movements in analogical and iconic form. But it is the ants that reflect the most complex social organization and display extraordinary related behavior. The farming leaf-cutter ants of the Americas are said to be the most advanced. Colonies of *Atta* comprise more than a million individuals, subdivided into numerous specialized worker castes, which coordinate their work through a multimodal communication system. Foragers move along a road network maintained by road-worker crews, cutting and transporting leaf fragments back to garden plots, where they are planted to produce fungus, which subsequently is harvested to feed the colony. The quantity of sediment moved to excavate one of their nest complexes in Brazil was 40 tons.[2]

Naturalists and philosophers have long marveled at the human-like features of insect societies, but the most remarkable aspect of the parallel is that the characteristics of eusociality should have surfaced among a hominoid taxon. With their massive bodies, low population densities, slow reproductive rates, and protracted infant dependences, apes and humans are in many ways the antithesis of the social insects. Eusociality is extremely rare among vertebrates, undoubtedly due to their reproductive biology. The only examples are found among the relatively fast-breeding rodents, specifically two genera of African mole rat.[3] As a result of some improbable developments in their evolutionary history, humans ended up re-creating features of the social insects, which provide an evolutionary context for humans as important as that of the primates.

The historical or genetic evolutionary relationship between social insects and humans is obviously remote. Humans belong to the phylum Chordata, which separates them from the insects by hundreds of millions of years. The similarities between the two are a classic case of *convergent evolution*,[4] in which the same or similar features evolved independently in unrelated groups. In this case, the convergence seems to reflect the emergent properties of highly complex systems.[5] Among the eusocial insects, these properties yield a super-organism.[6]

Although modern humans are rapidly approaching the point at which they could reproduce themselves biotechnologically in a manner similar to that of the insects, the complex social and economic hierarchies of their organizations are not based on reproductive biology and genetic relationships. They are instead based on information or, rather—because genes themselves are a form of information—a type of information that is coded symbolically in vast networks of nerve cells in the brain. Insects also store coded information in their brains, but if the organizational adaptations of humans are constrained by their reproductive biology, the storage and manipulation of nongenetic information among the insects is constrained by their small body and brain size. Humans inherited a large brain from the African apes, and it tripled in size within a few million years. Moreover, at a critical point in their later evolution, humans developed the structures (for example, an expanded prefrontal cortex) that are necessary to integrate many brains into one—a super-brain.

Humans evolved two features related to the coded information stored in neural networks that rendered them unique among all living organisms.

First, they developed the ability to translate that information into various material forms outside the brain—to create phenotypes of ideas rather than genes. Second, they developed the ability to combine and recombine that information with a sufficient number of elements and hierarchical levels that the variety of creations is potentially infinite.

After the formation of the super-brain among anatomically modern humans in Africa roughly 100,000 years ago, human organizations came to exhibit properties not previously observed in organic evolution. They created a growing and ultimately immense quantity of structured information outside the brain, and the processes by which the structures were generated and selected cannot be accounted for by the principles of evolutionary biology (genetic mutation and natural selection). They accumulated knowledge—organized nongenetic information—as a super-organismal entity, unconstrained by biological space or time.

These human organizations or societies were not super-organisms, however. The reproductive biology never caught up with the neocortical integration, yielding a society largely composed of competing and cantankerous individuals without close genetic ties. The result is a freakish variant of eusociality in which the collective mind—the super-brain—coexists uncomfortably with a group of organisms.

The Mind as History

[A]ll progress depends on the unreasonable man.
GEORGE BERNARD SHAW

Humans are odd creatures. They are organisms produced by the process of biological evolution, but they also participate in the phenomenon of the mind. They lead conflicted lives.

On the one hand, humans cannot escape the realities of organic life; consciousness itself is dependent on the functioning of evolved systems for the digestion of food, circulation and oxygenation of blood, release of neurotransmitters in the brain, and other bodily functions. Much of human life revolves around issues clearly related to evolutionary biology: the acquisition of energy, control of territory, defense of resources, reproduction and provisioning of offspring, and aging and death. On the other hand, humans receive and generate mental representations as part of a collective entity that seems to operate in accordance with the emergent properties of systems more complex than those previously observed in organic evolution. The structures of the super-brain or mind are generated (and selected) by a process of recombining information in complex hierarchical form that is unknown among other organisms.

I began with a discussion of how scholars of the seventeenth century addressed the problem of the mind. They were attempting to explain the universe in mechanistic terms and to place humans somewhere in this technologically inspired view of reality. Descartes laid the foundation for a debate that continues today by excluding the mind from natural science. There was opposition to this view from the outset, for example from Thomas Hobbes, and it only increased as technology and scientific explanation continued to progress through the eighteenth and nineteenth centuries.

The discovery of a mechanism for evolutionary change in the mid-nineteenth century was a major development. Darwin's idea was so powerful that it seems to have tipped the balance toward incorporating humanity into natural science. It significantly reinforced a trend that already was apparent in the writings of Johann Herder, Herbert Spencer, and others.[7] Thus in the early twentieth century, the philosopher of history R. G. Collingwood found little support for his complaint against the "tyranny of natural science" over philosophy and his contention that history is an autonomous form of thought—as well as the key to understanding the mind.[8] One of his allies was V. Gordon Childe. But Childe had been pigeon-holed as a doctrinaire Marxist by the 1940s, and his later writings were largely ignored.[9]

Reviewing centuries of thought about history, Collingwood praised those who had affirmed the autonomy of mind and history. He was particularly impressed with the work of his contemporary Benedetto Croce (1866-1952) in Italy.[10] Collingwood also admired the historical thinking of the two most eminent German philosophers: Immanuel Kant (1724-1804) and G. W. F. Hegel (1770-1831). Both had an advantage over many of their successors: they felt no compulsion to explain the mind in the context of evolutionary biology.

Kant was not a historian and wrote only a few short pieces on the subject, but his observations and reflections are still worth reading. He was appalled by the senseless waste and folly that characterized much of history and referred disparagingly to the "idiotic course of things human."[11] But like others of his epoch, he accepted the idea that history was progress, if somewhat halting and uneven. Kant defined this progress as the development of the mind and emphasized the cumulative nature of the process. According to Collingwood, "he identified the essence of mind as . . . autonomy, the power to make laws for oneself. This enabled him to put forward a new interpretation of the idea of history as the education of the human race."[12] Kant went on to address the question of what drives progress. His answer was "antagonisms" or conflicts within and between societies. Because individuals constantly put their selfish desires above the common good, Kant concluded that "from such crooked wood as man is made of, nothing perfectly straight can be built." Moreover, much of the attention of nation-states was consumed by war or "constant readiness for war." In their efforts to overcome these evils, Kant believed, humans had made and would continue to make progress toward a more rational world.[13]

Although notoriously obtuse, Hegel articulated similar ideas about both the direction and the propulsion of history. These ideas, which reflected the unmistakable influence of Kant, were presented in the introduction to his *Philosophy of History* (1840). Hegel also endorsed the notion of progress in history ("an advance to something better, more perfect"), which he felt was lacking in nature ("only a perpetually self-repeating cycle").[14] He saw the passion of individuals as the driving force behind change and progress because "in history an additional result is commonly produced by human actions beyond that which they aim at and obtain. . . . They gratify their own interest; but something further is thereby accomplished, latent in the actions in question, though not present to their consciousness, and not included in their design."[15] Even more so than Kant, Hegel confined his view of historical progress to the political realm, rather than the broader growth of knowledge, and this, as Collingwood noted, encouraged him to the unfortunate conclusion that it had achieved culmination in the eighteenth-century nation-state of Prussia.[16]

The conflict that Kant and Hegel identified as the catalyst for historical progress is the dissonance between individual humans as organisms enhancing their own reproductive fitness and the organizational creations of the collective mind. It is the same distinction made earlier that modern human society is based largely on a super-brain, not a super-organism. The genetics of the super-organism yield a society composed of individuals that function like the organs or cells of an organism; everyone follows the program, and there are no runoff elections or letters to the editor. The super-brain produces something very different.

Progress in the form of technological innovation with widespread social and economic consequences is especially evident in capitalist industrial societies after 1500. Much of the creation and dissemination of novel technologies has been driven by competition in the commercial sector. The novelties cover everything from jumbo jets to disposable diapers. Some of them, such as the mass-produced automobile or the birth control pill, have clearly effected changes large and small on the societies and super-brains that made them. It is more difficult however, to see how competition among individuals in earlier or simpler societies has driven innovation. Was this the source of writing or the spear-thrower?

Competition between societies—often erupting into open conflict—is another inescapable source of innovation. Here, too, the best examples derive from nation-states of the industrial era (although not necessarily capitalist nations). International competition drove innovation in military technology after 1400 across Europe, and the process became especially dynamic after the mid-nineteenth century. But in these cases, the antagonists represent separate super-brains, or at least poorly integrated components of a super-brain (for example, the French and Germans), and the dissonance between individual organisms and the collective mind would not seem to apply. In fact, it would seem to be the reverse: patriotic individuals laboring on the design of armored vehicles or microwave radar for the homeland appear to function as parts of a super-organism after all. The same principle is illustrated by individual sacrifices on the battlefield.[17] Moreover, some of the technological innovations, particularly in the realm of communications, have been steadily enhancing the integration of the super-brain, including the integration of those collective minds that once represented antagonistic nation-states. The most striking example is Europe in the early twenty-first century, but the broader trend is global. The diversification of super-brains that began more than 50,000 years ago in Africa with the dispersal of modern humans has been reversed. Humans seem destined to re-create the original unified super-brain or mind of the later African Middle Stone Age.

Perhaps the critical element in historical progress is the evolved capacity of the mind to imagine the future, to create alternative realities. It may be recalled that the essential function of the metazoan brain is to acquire information about changes in the environment and that the creative powers of the human mind represent a unique variant of this function—imagining things that never were and asking why not. The imaginings range from an altered shape of an arrow point to the workers' paradise and include an immense body of idiocy, but they clearly provide a constant stimulus for actions that may have significant consequences. Other animals may respond to the visual image of a food item or the ominous tread of a predator, but only humans risk their lives to pursue daydreams.

A History of the Future

[Y]ou can make systems that think a million times faster than a person. With AI, these systems could do engineering design. Combining this with the capability of a system to build something that is better than it, you have the possibility for a very abrupt transition.

K. ERIC DREXLER

It was Lord Byron who suggested that he and each of the guests at his villa on Lake Geneva in the summer of 1816 write "a ghost story," as recounted years later by Mary Shelley. The future wife of Percy Bysshe Shelley eventually composed a tale inspired by discussions between Shelley and Byron about the principle of life and the experiments of Erasmus Darwin. She began to wonder if "perhaps the component parts of a creature might be manufactured" and endowed with life.[18]

Ironically, the tale that Mary Shelley wrote and later published as a novel was the antithesis of a ghost story. *Frankenstein* is not about the returning spirit of a dead person, but about a human produced by technology. The genre is science fiction, and it may be considered a literary variant of the speculations of Francis Bacon. Writing two centuries later, Shelley could see, quite accurately it seems, where nineteenth-century technological progress was leading. Her novel was a deeply religious work that reflected not the bright optimism of Bacon, but the conviction that the inevitable attempt to "mock the stupendous mechanism of the Creator" would produce a monster.[19]

Like so many projects, from agriculture to pyramids, this one has deep roots in the Upper Paleolithic. In one way or another, we have been working on the creation of life and mind for at least 50,000 years—in both robotics and biotechnology. The earliest robots were the traps and snares designed possibly as early as 40,000 years ago and almost certainly by the later Upper Paleolithic. These were simple devices built to perform a single and highly specialized human function: catching a small mammal or bird (aquatic versions were designed for fish). They were "smart machines" with a cognitive function: respond to a stimulus received from the prey animal as it stepped into the snare or trap. In theory, they could be produced in large numbers and deployed across wide areas along traplines, creating an army of robots to perform important tasks for a small group of humans.

More broadly speaking, mechanical technology, which is unique to modern humans and the mind, mimics the functions of an organism. At first, mechanical technology operated as an extension of the body, drawing its energy entirely from human muscle power. The creation of technologies powered by other sources, initially wind and moving water, was a major step toward building a machine that could mimic a whole organism. The mechanical clock, although ultimately powered by muscle, stored that energy and controlled its release over an extended period of time—like a beating heart—through an ingenious escapement device. Clocks fascinated people: Thomas Hobbes referred to them as "artificiall life." They had a profound effect on how their makers viewed the world, including the living world.

As to the manufacture of the "component parts" that Mary Shelley mentioned, here also the history is lengthy. It has a precursor in the Upper Paleolithic with the production of various technologies designed to enhance the function of specific body parts: tailored clothing to function as artificial skin and hair, a spear-thrower to serve as an elongated arm, and artificial memory systems to enhance brain function. It began in earnest with the manufacture of replacement parts—prosthetic limbs, wooden teeth, and wigs, among others—which apparently dates to at least 500 B.C.E. and had became common by the sixteenth century in Europe. In recent decades, impressive advances have been made with the application of electronic information technology to artificial organ function, including vision.

Key elements of the brain also have been replicated technologically. The storage of information in coded digital form dates to the early Upper Paleolithic, and with exponential increases in the size and complexity of human interaction networks in the postglacial epoch, the quantity of data stored on information technology became immense. The computational function of the brain also was re-created, at first with comparatively simple mechanical devices like the ancient Greek Antikythera Mechanism and Charles Babbage's analytical engine of 1837.[20] Powerful computing machines awaited developments in electronics that yielded programmable digital computers in the 1940s that greatly exceeded the computational abilities of the human brain. At the same time, other technologies have replicated another key function of the animal brain: receiving sensory input in various media (chemical, tactile, sound, and light) and transforming it into coded data sets. These devices have, in turn, often been connected to a computing machine to analyze the received data—for example, a radar early-warning system that detects fast-moving objects and identifies potential incoming missiles.

For biotechnology, the Upper Paleolithic background extends back at least to the millennia following the maximum cold of the last glacial (18,000 years ago). The genetic modification of canids was succeeded by the domestication of various plants and animals in the postglacial epoch, and—as during the Upper Paleolithic—progress was tied to settlement and reduced mobility. And as with the mechanical clock, the technology provided insights into natural processes; the origin and evolution of life eventually was incorporated into the same mechanistic worldview. Recent advances in biotechnology have, of course, been extraordinary, and it is now possible to manipulate genetic material directly to produce desired phenotypes (for example, genetically modified corn).

At the beginning of the twenty-first century, it is apparent that the issue is not the creation of life (although this is a concern for many people). It is the creation of a mind, or true artificial intelligence ("strong AI"), that represents the most profound potential step. After 50,000 years of growth and development, the mind has achieved a substantial understanding of and some measure of control over organic life. We can effectively imitate life, including a higher animal brain, with robotics and information technology, while genetic engineering provides opportunities for rearranging organic forms. It is the generative properties of the mind that remain mysterious and thus far impossible to replicate with technology. Perhaps these properties will be comprehended only if and when AI can be successfully engineered, which, in turn, will probably shed light on the processes and events that underlay the emergence of the mind in anatomically modern humans.[21]

But how is it possible to build a machine that might, for example, write a novel like *Frankenstein?* Alan Turing, universally regarded as the founder of modern computing science, famously stated that a truly intelligent machine would have to be able to speak or write like a human.[22] AI has yet to be created, and some have argued that it is impossible. For one thing, it implies mechanical consciousness, which is hard for most people to imagine. Skepticism increased after earlier forecasts of an imminent breakthrough proved too optimistic, although some continue to predict that AI is coming; for example, the date proposed by Ray Kurzweil is 2045.[23] If so, perhaps it will be achieved, as was the digital computer, only after new developments in other areas, such as nanotechnology, are applied to the problem.[24]

Because AI seems remote and quite possibly unattainable, it has not been an issue of much concern. Far more attention has been paid to recent developments in biotechnology, such as genetically modified foods and cloning of animals, and the current and likely future directions of this applied science. A small number of people have nevertheless questioned the potential consequences of AI. A concern is that AI could be used for unscrupulous purposes as an ultimate weapon of sorts.[25] But another worry is that AI could, and perhaps inevitably would, become an independent entity—a mind that, in Kurzweil's words, "transcends biology." Supposing the mind could free itself from its host organisms and exist in an entirely nonbiological form?[26]

To begin with, the mind would have to retain or acquire two characteristics from its biological antecedents, rather like Mary Shelley's creature. The first is a pair of hands or some other means to effect changes in its environment. Otherwise, it would be reduced to persuading humans to translate its thoughts into action. This includes the means to re-engineer itself: to continue progressive development. The second is motivation: the source of historical progress that Hegel identified as "passion" and Shaw described as "unreasonable." Humans would have to create AI in their own image with the mechanical equivalent of hormones.

What would be the impact on humanity of an entity with vastly superior computational and perhaps even greater creative powers than our own? The mind could be liberated from the biological processes that produced it. It would require an energy source, but no food, water, oxygen, or sleep; no clothing, shelter, bed, or toilet. Among other things, the mind could embark on serious space exploration; unconstrained by biological time or the need to sustain an earthly environment, a mind could drift across the galaxy and beyond for millions of years. What would be the role of humans in such a world or, for that matter, of any life-forms? What purpose would they serve? Would they not be like rats in the cellar?

In a remote future, the mind might look back on its own history in the same way that humans reconstruct and explain their evolution from less complex forms of life. It might be a little embarrassing, much as "ape" ancestry was for the Victorians who contemplated it in Charles Darwin's time. What place would humanity occupy in the mind's heroic account of its own origin? We might be viewed as simply the final, and relatively brief, transitional phase of a long process that began with the first information systems more than 500 million years ago. From this perspective, *Homo* would appear as a handy vehicle through which one of these information systems could assemble itself into the mind, before moving on to bigger and better things.

NOTES

1. Modernity and Infinity

1. Vladimir Nabokov, *Speak, Memory: An Autobiography Revisited* (New York: Vintage, 1967), 297.
2. Vladimir Nabokov, *Lolita* (New York: Putnam, 1955).
3. Wild cattle or aurochs (*Bos primigenius*) are often depicted in cave paintings of Upper Paleolithic age in Europe, and the earliest of these have been directly dated to around 35,000 calibrated radiocarbon years ago. See, for example, Jean Clottes, *Chauvet Cave: The Art of Earliest Times* (Salt Lake City: University of Utah Press, 2003), 187.
4. Kathy Schick and Nicholas Toth, *Making Silent Stones Speak: Human Evolution and the Dawn of Technology* (New York: Simon and Schuster, 1993), 231–260; Thomas Wynn, "Handaxe Enigmas," *World Archaeology* 27 (1995): 10–24. On the dating of the earliest known bifacial stone tools, see Berhane Asfaw et al., "The Earliest Acheulean from Konso-Gardula," *Nature* 360 (1992): 732–735.
5. A rare and famous exception to this is the "dance" of the honeybee, which is performed to communicate information about food sources and potential nest sites to other bees. See Karl von Frisch, *The Dance Language and Orientation of Bees* (Cambridge, Mass.: Harvard University Press, 1967); and Thomas D. Seeley, *Honeybee Democracy* (Princeton, N.J.: Princeton University Press, 2010). See also Derek Bickerton, *Language and Human Behavior* (Seattle: University of Washington Press, 1995), 12–18.
6. For a definition and description of mental representation, see, for example, Daniel C. Dennett, *Brainstorms: Philosophical Essays on Mind and Psychology*

(Cambridge, Mass.: MIT Press, 1978), 41–50; Fred Dretske, *Explaining Behavior: Reasons in a World of Causes* (Cambridge, Mass.: MIT Press, 1988), 51–77; Hilary Putnam, *Representation and Reality* (Cambridge, Mass.: MIT Press, 1988), 37–41; and Steven Pinker, *How the Mind Works* (New York: Norton, 1997), 79–88.

7. See, for example, Richard G. Klein, *The Human Career: Human Biological and Cultural Origins*, 3rd ed. (Chicago: University of Chicago Press, 2009), 271–278.

8. The best general reference on the evolution of the human hand is John Napier, *Hands*, rev. Russell H. Tuttle (Princeton, N.J.: Princeton University Press, 1993).

9. Leslie Aiello and Christopher Dean, *An Introduction to Human Evolutionary Anatomy* (London: Academic Press, 1990), 232–243.

10. Ibid., 232.

11. Schick and Toth, *Making Silent Stones Speak*, 135–140.

12. Ibid., 231–245.

13. John F. Hoffecker, "Representation and Recursion in the Archaeological Record," *Journal of Archaeological Method and Theory* 14 (2007): 370–375.

14. Ibid., 375–379.

15. Lynn Wadley, "What Is Cultural Modernity? A General View and a South African Perspective from Rose Cottage Cave," *Cambridge Archaeological Journal* 1 (2001): 201–221.

16. Klein, *Human Career*, 690.

17. Christopher Henshilwood and Curtis W. Marean, "The Origin of Modern Human Behavior: Critique of the Models and Their Test Implications," *Current Anthropology* 44 (2003): 627–651.

18. Wadley, "What Is Cultural Modernity," 207–208.

19. Noam Chomsky, *On Nature and Language* (Cambridge: Cambridge University Press, 2002), 45; Marc D. Hauser, Noam Chomsky, and W. Tecumseh Fitch, "The Faculty of Language: What Is It, Who Has It, and How Did It Evolve?" *Science* 298 (2002): 1569–1579.

20. Noam Chomsky, *Language and the Problems of Knowledge: The Managua Lectures* (Cambridge, Mass.: MIT Press, 1988), 169–170. Chomsky also has applied the terms "recursion" and "merge" to the same phenomenon in *Language and Mind*, 3rd ed. (Cambridge: Cambridge University Press, 2006), 183–184.

21. A recent trend in the study of language (the "principles-and-parameters model" and associated themes) represents a departure from traditional concerns with "rules" of syntax and a move toward how language is related to "systems that are internal to the mind." Noam Chomsky wrote that "languages have no rules in anything like the familiar sense, and no theoretically significant grammatical constructions except as taxonomic artifacts" (*The Minimalist Program* [Cambridge, Mass.: MIT Press, 1995], 5–6).

22. W. Tecumseh Fitch and Marc D. Hauser, "Computational Constraints on Syntactic Processing in a Nonhuman Primate," *Science* 303 (2004): 377–380.

23. Hauser, Chomsky, and Fitch wrote that "the human faculty of language appears to be organized like the genetic code—hierarchical, generative, recursive, and virtually limitless with respect to its scope of expression" ("Faculty of Language," 1569).

24. Michael Corballis, "Recursion as the Key to the Human Mind," in *From Mating to Mentality: Evaluating Evolutionary Psychology*, ed. Kim Sterelny and Julie Fitness (New York: Psychology Press, 2003), 155-171.

25. Sarnoff A. Mednick, "The Associative Basis of the Creative Process," *Psychological Review* (1962): 220-232, quoted in Andreas Kyriacou, "Innovation and Creativity: A Neuropsychological Perspective," in *Cognitive Archaeology and Human Evolution*, ed. Sophie A. de Beaune, Frederick L. Coolidge, and Thomas Wynn (Cambridge: Cambridge University Press, 2009), 15-24.

26. Margaret A. Boden, *The Creative Mind: Myths and Mechanisms*, 2nd ed. (London: Routledge, 2004), 3-6.

27. Hoffecker, "Representation and Recursion in the Archaeological Record," 367-376.

28. Ibid., 377-378.

29. Wind instruments ("pipes") have been recovered from early Upper Paleolithic sites at Hohle Fels Cave and Giessenklosterle (both Germany) that date to more than 30,000 years ago. See Francesco d'Errico et al., "Archaeological Evidence for the Emergence of Language, Symbolism, and Music—An Alternative Multidisciplinary Perspective," *Journal of World Prehistory* 17 (2003): 1-70; and Nicholas J. Conard, Maria Malina, and Susanne C. Münzel, "New Flutes Document the Earliest Musical Tradition in Southwestern Germany," *Nature* 460 (2009): 727-740.

30. For a summary review of technological innovations in northern Eurasia during this period, see John F. Hoffecker, "Innovation and Technological Knowledge in the Upper Paleolithic of Northern Eurasia," *Evolutionary Anthropology* 14 (2005): 186-198.

31. For definitions of "symbol" and "icon," originally proposed by Charles Sanders Peirce in 1867, see, for example, Raymond Firth, *Symbols: Public and Private* (Ithaca, N.Y.: Cornell University Press, 1973), 60-62.

32. Francesco d'Errico, "Palaeolithic Origins of Artificial Memory Systems: An Evolutionary Perspective," in *Cognition and Material Culture: The Archaeology of Symbolic Storage*, ed. Colin Renfrew and Chris Scarre (Cambridge: McDonald Institute for Archaeological Research, 1998), 19-50. The suggestion that some of these artifacts might be lunar calendars was made by Alexander Marshack, *The Roots of Civilization: The Cognitive Beginnings of Man's First Art, Symbol and Notation* (New York: McGraw-Hill, 1972).

33. For an account of the development of the mechanical clock, see David S. Landes, *Revolution in Time: Clocks and the Making of the Modern World* (Cambridge, Mass.: Harvard University Press, 1983).

34. Lewis Mumford, *Technics and Civilization* (New York: Harcourt, Brace, 1934). Many other authors have commented on the impact of the clock on European worldview, including Daniel J. Boorstin, *The Discoverers: A History of Man's Search to Know His World and Himself* (New York: Random House, 1983), 26-78; and Donald Cardwell, *The Norton History of Technology* (New York: Norton, 1995), 37-101.

35. René Descartes, *Philosophical Works*, trans. Elizabeth S. Haldane and G. R. T. Ross (New York: Dover, 1955).

36. See, for example, Jaegwon Kim, *Philosophy of Mind*, 2nd ed. (Boulder, Colo.: Westview Press, 2006); and Keith T. Maslin, *An Introduction to the Philosophy of Mind*, 2nd ed. (Cambridge: Polity, 2007).

37. Hobbes's comments on Descartes's mind–body dualism were published in an appendix to *Six Metaphysical Meditations* in 1641. See Gilbert Ryle, *The Concept of Mind* (New York: Barnes & Noble, 1949).

38. Ernst Mayr, *The Growth of Biological Thought: Diversity, Evolution, and Inheritance* (Cambridge, Mass.: Harvard University Press, 1982).

39. Charles Darwin, "Variation Under Domestication," in *On the Origin of Species* (London: John Murray, 1859).

40. Ernst Mayr and William B. Provine, eds., The *Evolutionary Synthesis: Perspectives on the Unification of Biology* (Cambridge, Mass.: Harvard University Press, 1980); Mayr, *Growth of Biological Thought*.

41. Darwin wrote that "there is no fundamental difference between man and the higher mammals in their mental faculties" (*The Descent of Man and Selection in Relation to Sex* [London: John Murray, 1871], 66). The contrasting views of Darwin and Wallace on the mind are discussed in Chomsky, *Language and Mind*, 173–185.

42. Pinker, *How the Mind Works*, 521.

43. The technology most likely to offer insight into the workings of the mind has not been developed, however. It is not the clock, the cow, or the computer, but *artificial intelligence*, or AI. Although the computational theory of the mind accounts for the metazoan brain—including humans up through the australopithecines and perhaps early *Homo*—it accounts for only part of the modern human mind. If "strong AI" can be engineered in the next few decades (some are skeptical that it can ever be done), we will, by definition, understand the principles that govern creativity and discrete infinity. See, for example, Ray Kurzweil, *The Singularity Is Near: When Humans Transcend Biology* (New York: Viking Press, 2005).

44. Pinker, *How the Mind Works*, 525. See also Jerome H. Barkow, Leda Cosmides, and John Tooby, eds., *The Adapted Mind: Evolutionary Psychology and the Generation of Culture* (New York: Oxford University Press, 1992). For a critique of the computational theory of mind, see Jerry A. Fodor, *The Mind Doesn't Work That Way: Scope and Limits of Computational Psychology* (Cambridge, Mass.: MIT Press, 2000).

45. John Morgan Allman, *Evolving Brains* (New York: Scientific American Library, 1999); Larry W. Swanson, *Brain Architecture: Understanding the Basic Plan* (Oxford: Oxford University Press, 2003). For symbols in the brain, see Derek Bickerton, *Language and Species* (Chicago: University of Chicago Press, 1990), 77–87.

46. Richard Dawkins, *The Selfish Gene* (New York: Oxford University Press, 1976), 203–215.

47. The notion that the unique properties of the mind emerge from a highly complex system is sometimes described as *emergent materialism* and has been articulated by the philosopher Mario Bunge in *The Mind–Body Problem: A Psychobiological Approach* (Oxford: Pergamon Press, 1980), and *Emergence and Convergence: Qualitative Novelty and the Unity of Knowledge* (Toronto: University of Toronto Press, 2003). See also Vernon B. Mountcastle, *Perceptual Neuroscience: The Cerebral Cortex* (Cambridge, Mass.: Harvard University Press, 1998), 14–16. Chomsky

noted that the problem posed by Descartes more than 300 years ago remains unaddressed and that there is still "no substantive 'body of doctrine' about the ordinary creative use of language and other manifestations" of creativity (*On Nature and Language*, 55).

48. Gerald M. Edelman, *Wider Than the Sky: The Phenomenal Gift of Consciousness* (New Haven, Conn.: Yale University Press, 2004), 14–19. Elkhonon Goldberg also wrote, "The human brain is the most complex natural system in the known universe" (*The New Executive Brain: Frontal Lobes in a Complex World* [Oxford: Oxford University Press, 2009], 21). See also, for example, Marvin Minsky, *The Society of Mind* (New York: Simon and Schuster, 1986); and Steven Pinker and Jacques Mehler, *Connections and Symbols* (Cambridge, Mass.: MIT Press, 1988).

49. Merlin Donald, *Origins of the Modern Mind: Three Stages in the Evolution of Culture and Cognition* (Cambridge, Mass.: Harvard University Press, 1991). Throughout *Landscape of the Mind*, I have asserted that the mind "transcends biological space and time." In this case, I am defining biological space and time in terms of individual organisms and not genomes, which are biological entities that also transcend the space and time of individual organisms.

50. The dispersal of modern humans within and beyond Africa is best illustrated by genetic data collected from living humans. See Spencer Wells, *Deep Ancestry: Inside the Genographic Project* (Washington, D.C.: National Geographic Society, 2007).

51. Thomas Hobbes, *Leviathan, or the Matter, Forme and Power of a Common-wealth Ecclesiasticall and Civill* (London: Penguin, 1968), 81.

52. William Morton Wheeler, "The Ant-Colony as an Organism," *Journal of Morphology* 22 (1911): 307–325.

53. Edward O. Wilson, *The Insect Societies* (Cambridge, Mass.: Belknap Press of Harvard University Press, 1971); Bert Hölldobler and E. O. Wilson, *The Super-Organism: The Beauty, Elegance, and Strangeness of Insect Societies* (New York: Norton, 2009), 10–13.

54. Hobbes, *Leviathan*, 226.

55. Darwin, *Origin of Species*, 237.

56. Hölldobler and Wilson, *Super-Organism*, 24–29. See also Dawkins, *Selfish Gene*.

57. Although reproductive biology would seem to confine eusociality to invertebrates, one of the few known examples outside the ants, bees, and termites is found among two genera of African mole-rat.

58. Richard C. Lewontin, *The Genetic Basis of Evolutionary Change* (New York: Columbia University Press, 1974). For a recent discussion of the relevance of Lewontin's findings for the dispersal of modern humans, see Wells, *Deep Ancestry*, 19–22.

59. Goldberg, *New Executive Brain*, 89–114.

60. Robin Dunbar, *Grooming, Gossip, and the Evolution of Language* (Cambridge, Mass.: Harvard University Press, 1996), and "The Social Brain Hypothesis," *Evolutionary Anthropology* 6 (1998): 178–190.

61. Gerald M. Edelman and Giulio Tononi, *A Universe of Consciousness: How Matter Becomes Imagination* (New York: Basic Books, 2000), 103.

62. Ibid., 193–199. Edelman and Tononi also wrote, "Once higher-order consciousness begins to emerge, a self can be constructed from social and affective

relationships. This self . . . goes far beyond the biologically based individuality of an animal with primary consciousness" (193).

63. Carl Mitcham, *Thinking Through Technology: The Path Between Engineering and Philosophy* (Chicago: University of Chicago Press, 1994), 20-24.

64. Ibid., 29-33. See also Carl Mitcham and Robert Mackey, "Introduction: Technology as a Philosophical Problem," in *Philosophy and Technology: Readings in the Philosophical Problems of Technology*, ed. Carl Mitcham and Robert Mackey (New York: Free Press, 1983), 22-25.

65. Martin Heidegger, "The Question Concerning Technology," in *The Question Concerning Technology and Other Essays*, trans. William Lovitt (New York: Harper & Row, 1977), 5-30. The concept or theory of *embodied cognition*, developed in the field of psychology, is explicitly based on Heidegger's philosophy. See, for example, Michael L. Anderson, "Embodied Cognition: A Field Guide," *Artificial Intelligence* 149 (2003): 91-30. Anderson summed up one of his colleague's views thusly: "[R]epresentations are therefore 'sublimations' of bodily experience, possessed of content already, and not given content or form by an autonomous mind" (104). In my view, embodied cognition would not survive an encounter with the archaeological record, which contains abundant evidence for external representations that have no basis in bodily experience, beginning probably with bifaces and including much of the geometric urban "landscape of the mind."

66. Samuel Butler, "Darwin Among the Machines," in *The Shrewsbury Edition of the Works of Samuel Butler*, ed. Henry Festing Jones (London: Cape, 1923), 1:208-210. Butler's ideas about mechanical life (as well as those of Hobbes) are described in George B. Dyson's remarkable book *Darwin Among the Machines: The Evolution of Global Intelligence* (Reading, Mass.: Addison-Wesley, 1997).

67. Jane Goodall, *In the Shadow of Man* (London: Collins, 1971).

68. David L. Clarke, *Analytical Archaeology*, 2nd ed. (New York: Columbia University Press, 1978), 181.

69. Robert Pool, *Beyond Engineering: How Society Shapes Technology* (New York: Oxford University Press, 1997), 119.

70. Schick and Toth, *Making Silent Stones Speak*, 130-133.

71. Hoffecker, "Representation and Recursion in the Archaeological Record," 380-381.

72. See, for example, Cordelia Fine, ed., *The Britannica Guide to the Brain* (London: Constable and Robinson, 2008), 73-86.

73. Bruce G. Trigger, *A History of Archaeological Thought* (Cambridge: Cambridge University Press, 1989).

74. V. Gordon Childe, *Society and Knowledge: The Growth of Human Traditions* (New York: Harper, 1956), 1.

75. Heidegger, "Question Concerning Technology," 5.

76. See, for example, Trigger, *History of Archaeological Thought*, 264-328; and Ian Hodder, *Reading the Past: Current Approaches to Interpretation in Archaeology*, 2nd ed. (Cambridge: Cambridge University Press, 1991), 19-34.

77. See, for example, Ian Hodder, "Theoretical Archaeology: A Reactionary View,"

in *Symbolic and Structural Archaeology*, ed. Ian Hodder (Cambridge: Cambridge University Press, 1981), 12-13.

78. Colin Renfrew, *Towards an Archaeology of Mind* (Cambridge: Cambridge University Press, 1982). See also Colin Renfrew, "What Is Cognitive Archaeology?" *Cambridge Archaeological Journal* 3 (1993): 247-270.

79. Donald, *Origins of the Modern Mind*. Donald probably was unaware that similar ideas had been discussed several decades earlier by the French archaeologist André Leroi-Gourhan, who wrote that "la mémoire de l'homme est extériorisée et son contenant est la collectivité ethnique" (*Le Geste et la parole*, vol. 2, *La mémoire et les rythmes* [Paris: Albin Michel, 1965], 64). For a translation, see André Leroi-Gourhan, *Gesture and Speech*, trans. Anna Bostock Berger (Cambridge, Mass.: MIT Press, 1993).

80. Colin Renfrew and Chris Scarre, eds., *Cognition and Material Culture: The Archaeology of Symbolic Storage* (Cambridge: McDonald Institute for Archaeological Research, 1998); Elizabeth DeMarrais, Chris Gosden, and Colin Renfrew, eds., *Rethinking Materiality: The Engagement of Mind with the Material World* (Cambridge: McDonald Institute for Archaeological Research, 2004). See also de Beaune, Coolidge, and Wynn, eds., *Cognitive Archaeology and Human Evolution.*

81. See, for example, Thomas Wynn, "Tools, Grammar and the Archaeology of Cognition," *Cambridge Archaeological Journal* 1 (1991): 191-206; and Colin Renfrew, "Cognitive Archaeology," in *Archaeology: The Key Concepts*, ed. Colin Renfrew and Paul Bahn (Abingdon: Routledge, 2005), 41-45.

82. See, for example, Claude Lévi-Strauss, *Structural Anthropology*, trans. Claire Jacobson and Brooke Grundfest Schoepf (New York: Basic Books, 1963); Edmund Leach, *Claude Lévi-Strauss* (New York: Viking Press, 1970); and Jean Piaget, *Structuralism*, trans. Chaninah Maschler (New York: Harper & Row, 1970).

83. Nathan Schlanger, "The *Chaîne Opératoire*," in *Archaeology*, ed. Renfrew and Bahn, 25-31. See also Leroi-Gourhan, *Gesture and Speech*, 231-235.

84. Stanley H. Ambrose, "Paleolithic Technology and Human Evolution," *Science* 291 (2001): 1748-1753; Jacques Pelegrin, "Cognition and the Emergence of Language: A Contribution from Lithic Technology," in *Cognitive Archaeology and Human Evolution*, ed. de Beaune, Coolidge, and Wynn, 95-108.

85. Steven Mithen, *The Prehistory of the Mind: The Cognitive Origins of Art, Religion and Science* (London: Thames and Hudson, 1996).

86. Hoffecker, "Representation and Recursion in the Archaeological Record," 370-375.

87. Gordon W. Hewes, "A History of Speculation on the Relation Between Tools and Language," in *Tools, Language and Cognition in Human Evolution*, ed. Kathleen R. Gibson and Tim Ingold (Cambridge: Cambridge University Press, 1993), 20-31.

88. Ralph Holloway, "Culture: A *Human* Domain," *Current Anthropology* 10 (1969): 395-412.

89. James Deetz, *Invitation to Archaeology* (Garden City, N.Y.: Natural History Press, 1967), 81-101; Glynn Ll. Isaac, "Stages of Cultural Elaboration in the Pleistocene:

Possible Archaeological Indicators of the Development of Language Capabilities," *Annals of the New York Academy of Sciences* 280 (1976): 275-288.

90. Thomas Wynn, a leading cognitive archaeologist, has been consistently skeptical of the parallels between language and toolmaking, stressing the differences over the similarities in "Tools, Grammar and the Archaeology of Cognition," and "The Evolution of Tools and Symbolic Behavior," in *Handbook of Human Symbolic Evolution*, ed. Andrew Lock and Charles R. Peters (Oxford: Blackwell, 1999), 269-271. The "modular model" of the mind was initially presented in Jerry A. Fodor, *The Modularity of Mind* (Cambridge, Mass.: MIT Press, 1983). It has had an impact on evolutionary psychology; see, for example, Leda Cosmides and John Tooby, "Origins of Domain Specificity: The Evolution of Functional Organization," in *Mapping the Mind: Domain Specificity in Cognition and Culture*, ed. Lawrence A. Hirschfield and Susan A. Gelman (Cambridge: Cambridge University Press, 1994), 85-116. For a discussion of the modular model of the mind in the context of cognitive archaeology, see Mithen, *Prehistory of the Mind*, 37-60.

91. Patricia M. Greenfield, "Language, Tools, and Brain: The Ontogeny and Phylogeny of Hierarchically Organized Sequential Behavior," *Behavioral and Brain Sciences* 14 (1991): 531-595.

92. Goldberg wrote that "the mental representation of a thing is not modular. It is distributed, since its different sensory components are represented in different parts of the cortex" (*New Executive Brain*, 54).

93. Karen Emmorey, Sonya Mehta, and Thomas J. Grabowski, "The Neural Correlates of Sign Versus Word Production," *NeuroImage* 36 (2007): 202-208.

94. Klein, *Human Career*, 647-649; Frederick L. Coolidge and Thomas Wynn, "Working Memory, Its Executive Functions, and the Emergence of Modern Thinking," *Cambridge Archaeological Journal* 15 (2005): 5-26.

95. Wolfgang Enard et al., "Molecular Evolution of FOXP2, a Gene Involved in Speech and Language," *Nature* 418 (2002): 869-872.

2. Daydreams of the Lower Paleolithic

1. Thomas Wynn, "Piaget, Stone Tools and the Evolution of Human Intelligence," *World Archaeology* 17 (1985): 32-43; John A. J. Gowlett, "The Elements of Design Form in Acheulian Bifaces: Modes, Modalities, Rules and Language," in *Axe Age: Acheulian Tool-making from Quarry to Discard*, ed. Naama Goren-Inbar and Gonen Sharon (London: Equinox, 2006), 203-221. The term "mental template" was introduced by the American archaeologist James Deetz, *Invitation to Archaeology* (Garden City, N.Y.: Natural History Press, 1967), 45-49.

2. See, for example, Thomas Wynn, "The Evolution of Tools and Symbolic Behaviour," in *Handbook of Human Symbolic Evolution*, ed. Andrew Lock and Charles R. Peters (Oxford: Blackwell, 1999), 269.

3. Bifaces flaked from bone have been recovered from Lower Paleolithic sites in Italy and Germany. See Paola Villa, "Middle Pleistocene Prehistory in Southwestern Europe: The State of Our Knowledge and Ignorance," *Journal of Anthropological Research* 47 (1991): 193-217; and Ursula Mania, "The Utilisation of

Large Mammal Bones in Bilzingsleben—A Special Variant of Middle Pleistocene Man's Relationship to His Environment," *ERAUL* 62 (1995): 239-246.

4. See, for example, Philip V. Tobias, *The Brain in Hominid Evolution* (New York: Columbia University Press, 1971), 106-108; and Vernon B. Mountcastle, *Perceptual Neuroscience: The Cerebral Cortex* (Cambridge, Mass.: Harvard University Press, 1998), 20-23.

5. David Marr, *Vision: A Computational Investigation into the Human Representation and Processing of Visual Information* (San Francisco: Freeman, 1982).

6. Richard G. Klein, *The Human Career: Human Biological and Cultural Origins*, 3rd ed. (Chicago: University of Chicago Press, 2009), 103.

7. John Morgan Allman, *Evolving Brains* (New York: Scientific American Library, 1999), 2-3.

8. Ibid., 3-8. See also Larry W. Swanson, *Brain Architecture: Understanding the Basic Plan* (Oxford: Oxford University Press, 2003), 11-14. A seminal reference on the early evolution of the brain is G. H. Parker, *The Elementary Nervous System* (Philadelphia: Lippincott, 1919).

9. Thomas Suddendorf, Donna Rose Addis, and Michael C. Corballis, "Mental Time Travel and the Shaping of the Human Mind," *Philosophical Transactions of the Royal Society B* 364 (2009): 1317-1324.

10. Swanson, *Brain Architecture*, 16-26; Allman, *Evolving Brains*, 9-13. See also Richard Dawkins, *The Ancestor's Tale: A Pilgrimage to the Dawn of Evolution* (Boston: Houghton Mifflin, 2004), 463-476.

11. Derek Bickerton, *Language and Species* (Chicago: University of Chicago Press, 1990), 77-87.

12. Mario F. Wulliman, "Brain Phenotypes and Early Regulatory Genes: The *Bauplan* of the Metazoan Central Nervous System," in *Brain Evolution and Cognition*, ed. Gerhard Roth and Mario F. Wulliman (New York: Wiley, 2001), 11-40; Swanson, *Brain Architecture*, 29-38. A general source on invertebrate brains is Theodore Holmes Bullock and G. Adrian Horridge, *Structure and Function in the Nervous Systems of Invertebrates* (San Francisco: Freeman, 1965).

13. Allman, *Evolving Brains*, 68-73.

14. Matthew H. Nitecki, ed., *Evolutionary Innovations* (Chicago: University of Chicago Press, 1990).

15. C. Gans and G. Northcutt, "Neural Crest and the Origin of the Vertebrates: A New Head," *Science* 220 (1983): 268-274; Allman, *Evolving Brains*, 73-79; Swanson, *Brain Architecture*, 40-79.

16. Allman, *Evolving Brains*, 75-78.

17. Dawkins, *Ancestor's Tale*, 293-319.

18. Allman, *Evolving Brains*, 92-106.

19. Mountcastle, *Perceptual Neuroscience*, 19-25; Georg F. Striedter, *Principles of Brain Evolution* (Sunderland, Mass.: Sinauer, 2005), 255-296.

20. Glenn C. Conroy, *Primate Evolution* (New York: Norton, 1990).

21. Striedter, *Principles of Brain Evolution*, 303-304.

22. Robin Dunbar, "The Social-Brain Hypothesis," *Evolutionary Anthropology* 6 (1998): 178-190.

23. Allman, *Evolving Brains*, 122–157; Streidter, *Principles of Brain Evolution*, 305–310.
24. Allman, *Evolving Brains*, 143–146; Streidter, *Principles of Brain Evolution*, 301–305.
25. Robert Martin, *Primate Origins and Evolution: A Phylogenetic Reconstruction* (Princeton, N.J.: Princeton University Press, 1990). It has been noted that tree squirrels possess laterally placed eyes, which seems to undercut the arboreal theory; despite the placement of their eyes, however, there is some overlap in their fields of vision.
26. Matt Cartmill, "Rethinking Primate Origins," *Science* 184 (1974): 436–443.
27. Allman, *Evolving Brains*, 144–146.
28. W. E. Le Gros Clark, *History of the Primates* (Chicago: University of Chicago Press, 1959), 45–53.
29. Allman, *Evolving Brains*, 146–149.
30. See, for example, Steven Pinker, *How the Mind Works* (New York: Norton, 1997), 213–214.
31. Marr, *Vision*, 3–6.
32. Michael C. Corballis, *The Lopsided Ape: Evolution of the Generative Mind* (New York: Oxford University Press, 1991), 218–222.
33. Marr, *Vision*, 37, table 1-1.
34. In this context, music—especially the complex musical structures developed by composers like Mozart and Bach—becomes a particularly striking creative product of the mind. Most of the material structures generated by the mind appear to have a logical basis in (restructured) natural visual representations. But complex sound structures seem to lack such a "natural" basis. Natural sounds tend to blend and are difficult to represent spatially in the brain with precision; accordingly, the artificial sounds produced by musical instruments are to a significant degree structured temporally. See Daniel J. Levitin, *This Is Your Brain on Music: The Science of a Human Obsession* (New York: Penguin, 2006).
35. Conroy, *Primate Evolution*, 200–248; Klein, *Human Career*, 112–119.
36. For an excellent account of *Proconsul* aimed at the general reader, see Alan Walker and Pat Shipman, *The Ape in the Tree: An Intellectual and Natural History of Proconsul* (Cambridge, Mass.: Belknap Press of Harvard University Press, 2005). See also Chris Stringer and Peter Andrews, *The Complete World of Human Evolution* (New York: Thames and Hudson, 2005), 92–95.
37. Klein, *Human Career*, 119–127. Brain volume for *Dryopithecus* has been estimated on the basis of a cranium recovered from Rudabánya (Hungary).
38. Ian Tattersall, *The Fossil Trail: How We Know What We Think We Know About Human Evolution* (New York: Oxford University Press, 1995), 3–4.
39. Roger Lewin and Robert A. Foley, *Principles of Human Evolution* (Malden, Mass.: Blackwell, 2004), 210–212.
40. Russell H. Tuttle, *Apes of the World: Their Social Behavior, Communication, Mentality and Ecology* (Park Ridge, N.J.: Noyes, 1986); William McGrew, *The Cultured Chimpanzee: Reflections on Cultural Primatology* (Cambridge: Cambridge University Press, 2004). Recent analysis of 4.4-million-year-old fossils assigned to *Ardipithecus* seem to confirm that early humans "never knuckle-walked." See Tim D. White et al., "*Ardipithecus ramidus* and the Paleobiology of Early Hominids," *Science* 326 (2009): 75–86.

41. Charles Darwin, On the Origin of Species (London: John Murray, 1859), 279-311.

42. Niles Eldredge, Time Frames: The Evolution of Punctuated Equilibria (Princeton, N.J.: Princeton University Press, 1985).

43. Views regarding the cognitive powers of chimpanzees and the other great apes vary widely. A number of primatologists believe that apes evolved their own unique level of cognition—including "generativity, and cognitive fluidity as well as . . . mental representation of absent items, perspective taking, cooperative hunting, food sharing, and symbolic communication"—that provided an essential context for the emergence of the human mind. See Anne E. Russon and David R. Begun, "Evolutionary Origins of Great Ape Intelligence: An Integrated View," in The Evolution of Thought: Evolutionary Origins of Great Ape Intelligence, ed. Anne E. Russon and David R. Begun (Cambridge: Cambridge University Press, 2004), 353-368. A more cautious approach has been pursued by Daniel Povinelli and his students, who base their conclusions on a series of controlled experiments designed to reveal knowledge among apes through object manipulation and problem solving; they concluded that chimpanzees are incapable of forming concepts about entities that they cannot observe directly. See Daniel J. Povinelli, Folk Physics for Apes: The Chimpanzee's Theory of How the World Works (Oxford: Oxford University Press, 2000).

44. Peter S. Rodman and Henry M. McHenry, "Bioenergetics and the Origin of Hominid Bipedalism," American Journal of Physical Anthropology 52 (1980): 103-106.

45. Klein, Human Career, 271-278.

46. Lewin and Foley, Principles of Human Evolution, 241-253.

47. John H. Langdon, The Human Strategy: An Evolutionary Perspective on Human Anatomy (New York: Oxford University Press, 2005), 116-128.

48. Jonathan Kingdon, Lowly Origin: Where, When, and Why Our Ancestors First Stood Up (Princeton, N.J.: Princeton University Press, 2003), 118-147.

49. Michel Brunet et al., "A New Hominid from the Upper Miocene of Chad, Central Africa," Nature 418 (2002): 145-801. See also Bernard Wood, Human Evolution: A Very Short Introduction (Oxford: Oxford University Press, 2005), 63-65.

50. K. Galik et al., "External and Internal Morphology of the BAR 1002'00 Orrorin tugenensis Femur," Science 305 (2004): 1450-1453.

51. Wood, Human Evolution, 65-67.

52. Tim D. White, Gen Suwa, and Berhane Asfaw, "Australopithecus ramidus, a New Species of Early Hominid from Aramis, Ethiopia," Nature 371 (1994): 306-312. The classification of these fossils was subsequently revised to the genus Ardipithecus. See White et al., "Ardipithecus ramidus and the Paleobiology of Early Hominids," 80-81.

53. Meave G. Leakey et al., "New Four-Million-Year-Old Hominid Species from Kanapoi and Allia Bay, Kenya," Nature 376 (1995): 565-571.

54. Lewin and Foley, Principles of Human Evolution, 261-262.

55. Randall L. Susman, Jack T. Stern, and William J. Jungers, "Locomotor Adaptations in the Hadar Hominids," in Ancestors: The Hard Evidence, ed. Eric Delson (New York: Liss, 1985), 184-192.

56. There are four major references on the hand. The earliest is Charles Bell, *The Hand: Its Mechanism and Vital Endowments as Evincing Design* (New York: Harper, 1840). The second is John Napier, *Hands*, rev. Russell H. Tuttle (New York: Pantheon Books, 1980; Princeton, N.J.: Princeton University Press, 1993). The third is Frank R. Wilson, *The Hand: How Its Use Shapes the Brain, Language, and Human Culture* (New York: Pantheon Books, 1998). The most recent source, with emphasis on neuroanatomy, is Vernon B. Mountcastle, *The Sensory Hand: Neural Mechanisms of Somatic Sensation* (Cambridge, Mass.: Harvard University Press, 2005).

57. Mountcastle, *Sensory Hand*, 35–36.

58. Ibid., 41–44.

59. Ibid., 1–6.

60. Napier, *Hands*, 55.

61. The length of the thumb × 100/length of the index finger = the *opposability index*, which is 60 for living humans. See ibid., 57–58.

62. Ibid., 61–63.

63. John Napier, "The Evolution of the Hand," *Scientific American* 207 (1962): 56–62, and *Hands*, 29–38.

64. Napier, *Hands*, 62; Mountcastle, *Sensory Hand*, 38–40.

65. Napier, *Hands*, 73–74.

66. Ibid., 57–58.

67. Walker and Shipman, *Ape in the Tree*, 74–75.

68. Salvador Moyà-Solà and Meike Köhler, "A *Dryopithecus* Skeleton and the Origin of Great-Ape Locomotion," *Nature* 379 (1996): 156–159; Lewin and Foley, *Principles of Human Evolution*, 219–220.

69. Napier, *Hands*, 67–77.

70. White et al., "*Ardipithecus ramidus* and the Paleobiology of Early Hominids," 80–81.

71. Michael E. Bush et al., "Hominid Carpal, Metacarpal, and Phalangeal Bones Recovered from the Hadar Formation: 1974–1977 Collections," *American Journal of Physical Anthropology* 57 (1982): 651–677.

72. Mary W. Marzke, "Joint Functions and Grips of the *Australopithecus afarensis* Hand, with Special Reference to the Region of the Capitate," *Journal of Human Evolution* 12 (1983): 197–211. See also Leslie Aiello and Christopher Dean, *An Introduction to Human Evolutionary Anatomy* (London: Academic Press, 1990), 385–388.

73. John Napier, "Fossil Hand Bones from Olduvai Gorge," *Nature* 196 (1962): 409–411, and *Hands*, 88–89; Aiello and Dean, *Introduction to Human Evolutionary Anatomy*, 389–392. The hand bones include three carpals, eleven phalanges, and one metacarpal fragment from Olduvai Hominid 7 assigned to *Homo habilis* from Bed I at Olduvai Gorge.

74. Randall L. Susman, "Hand of *Paranthropus robustus* from Member 1, Swartkrans: Fossil Evidence for Tool Behavior," *Science* 240 (1988): 781–784.

75. Klein, *Human Career*, 259–261.

76. Alan Walker and Richard E. Leakey, "The Postcranial Bones," in *The Nariokotome Homo erectus Skeleton*, ed. Alan Walker and Richard E. Leakey (Berlin: Springer, 1993), 136–138. The bones include two first metacarpal shafts (?) and two phalanges.

77. John F. Hoffecker, "Representation and Recursion in the Archaeological Record," *Journal of Archaeological Method and Theory* 14 (2007): 359-387.

78. See, for example, Kathy Schick and Nicholas Toth, *Making Silent Stones Speak: Human Evolution and the Dawn of Technology* (New York: Simon and Schuster, 1993), 227-237. The term "Oldowan" is derived from Olduvai Gorge in Tanzania.

79. Sileshi Semaw, "The Oldest Stone Artifacts from Gona (2.6-2.5 Ma), Afar, Ethiopia: Implications for Understanding the Earliest Stages of Stone Knapping," in *The Oldowan: Case Studies into the Earliest Stone Age*, ed. Nicholas Toth and Kathy Schick (Gosport, Ind.: Stone Age Institute Press, 2006), 43-75. The term "Acheulean" is derived from Saint-Acheul in France.

80. See, for example, Mary D. Leakey, *Olduvai Gorge*, vol. 3, *Excavations in Beds I and II, 1960-1963* (Cambridge: Cambridge University Press, 1971). The volume describes and illustrates stone artifacts of Oldowan and Acheulean assemblages. Note that some artifacts classified as "proto-bifaces" of the Oldowan are fully bifacially flaked (for example, fig. 51[2] on p. 101), while some artifacts classified as "bifaces" of the succeeding Acheulean are only partially bifacially flaked (for example, fig. 63 on p. 129).

81. Bifaces do not appear outside Africa until about 1.4 million years ago, and they remain rare or absent in many parts of Eurasia during the Lower Paleolithic, especially in the Far East and eastern Europe. As far as we know, the economy of humans inhabiting these places (chiefly *Homo erectus*) was similar to that of humans living in Africa and western Europe, and they apparently used Oldowan tools to perform the same function(s) as large bifacial tools. See, for example, Noel T. Boaz and Russell L. Ciochon, *Dragon Bone Hill: An Ice-Age Saga of Homo erectus* (Oxford: Oxford University Press, 2004), 95-107. For the analysis of microscopic wear patterns on hand axes, see Lawrence H. Keeley, "Microwear Analysis of Lithics," in *The Lower Paleolithic Site at Hoxne, England*, ed. Ronald Singer, Bruce G. Gladfelter, and John J. Wymer (Chicago: University of Chicago Press, 1993), 129-138. Microscopic use-wear (almost entirely attributed to meat and bone polish from butchery) was identified on roughly forty bifaces from Boxgrove (England), as discussed in Michael Pitts and Mark Roberts, *Fairweather Eden: Life in Britain Half a Million Years Ago as Revealed by the Excavations at Boxgrove* (London: Century, 1998), 285-287.

82. Semaw, "Oldest Stone Artifacts from Gona," 68-69.

83. Kathy Schick and Nicholas Toth, "An Overview of the Oldowan Industrial Complex: The Sites and the Nature of Their Evidence," in *Oldowan*, ed. Toth and Schick, 18. Remains of *Homo* are associated with Oldowan artifacts at Hadar (Ethiopia), Olduvai Gorge (Tanzania), and Swartkrans (South Africa).

84. Nicholas Toth et al., "*Pan* the Toolmaker: Investigations into the Stone Tool-Making and Tool-Using Capabilities of a Bonobo (*Pan paniscus*)," *Journal of Archaeological Science* 20 (1993): 81-91; Nicholas Toth, Kathy Schick, and Sileshi Semaw, "A Comparative Study of the Stone Tool-Making Skills of *Pan*, *Australopithecus*, and *Homo sapiens*," in *Oldowan*, ed. Toth and Schick, 155-222.

85. Schick and Toth, "Overview of the Oldowan Industrial Complex," 28-29.

86. R. X. Zhu et al., "Early Evidence of the Genus *Homo* in East Asia," *Journal of Human Evolution* 55 (2008): 1075-1085.

87. Schick and Toth, "Overview of the Oldowan Industrial Complex," 9-10.

88. Leakey, Olduvai Gorge.

89. Nicholas Toth, "The Oldowan Reassessed: A Close Look at Early Stone Arti-facts," Journal of Archaeological Science 12 (1985): 101-120. See also Schick and Toth, "Overview of the Oldowan Industrial Complex," 5-9.

90. Schick and Toth, "Overview of the Oldowan Industrial Complex," 9-10; Klein, Human Career, 252-259.

91. See, for example, Richard Potts and Pat Shipman, "Cutmarks Made by Stone Tools on Bones from Olduvai Gorge, Tanzania," Nature 291 (1981): 577-580; and Henry T. Bunn and Ellen M. Kroll, "Systematic Butchery by Plio/Pleisto-cene Hominids at Olduvai Gorge, Tanzania," Current Anthropology 27 (1986): 431-452.

92. Lawrence H. Keeley and Nicholas Toth, "Microwear Polishes on Early Stone Tools from Koobi Fora, Kenya," Nature 293 (1981): 464-466.

93. Manuel Domínguez-Rodrigo and Travis Rayne Pickering, "Early Hominid Hunt-ing and Scavenging: A Zooarchaeological Review," Evolutionary Anthropology 12 (2003): 275-282.

94. Klein, Human Career, 157-160.

95. Potts and Shipman, "Cutmarks Made by Stone Tools on Bones from Olduvai Gorge, Tanzania."

96. Leakey, Olduvai Gorge, 1-3. Leakey defined the Acheulean assemblages at Oldu-vai Gorge as those containing at least 40 percent bifaces among the tools.

97. Berhane Asfaw et al., "The Earliest Acheulean from Konso-Gardula," Nature 360 (1992): 732-735; Klein, Human Career, 373-381.

98. Toth, "Oldowan Reassessed," 101-120.

99. Leakey, Olduvai Gorge, 21-123. The most likely candidate for an Oldowan arti-fact based on a mental template would seem to be those classified as "spheroids," which Leakey described as "stone balls, smoothly rounded over the whole exterior" (6). On the basis of experiments, Schick and Toth concluded that these probably represent heavily battered hammerstones: "[A]fter approximately four hours of percussion these quartz hammers assumed a remarkably spherical shape without any necessary intent or predetermination" (Making Silent Stones Speak, 130-133).

100. If there is some debate about the implications of other developments in Paleo-lithic technology for human cognitive faculties (for example, Levallois prepared cores), there is near consensus on the biface as a "mental template," including among archaeologists of the cognitive variety. See, for example, Steven Mithen, The Prehistory of the Mind: The Cognitive Origins of Art and Science (London: Thames and Hudson, 1996), 117-119; Jacques Pelegrin, "Cognition and the Emergence of Language: A Contribution from Lithic Technology," in Cognitive Archaeol-ogy and Human Evolution, ed. Sophie A. de Beaune, Frederick L. Coolidge, and Thomas Wynn (Cambridge: Cambridge University Press, 2009), 100-102; and Frederick L. Coolidge and Thomas Wynn, The Rise of Homo sapiens: The Evolution of Modern Thinking (Malden, Mass.: Wiley-Blackwell, 2009), 110-113. An alter-native view was expressed by William Noble and Iain Davidson: "The repeated forms of stone artefacts such as handaxes seem likely to be the result of learned

motor actions rather than deliberate design or 'planned' attempts to reproduce a 'mental template' of the ideal form" (*Human Evolution, Language and Mind: A Psychological and Archaeological Inquiry* [Cambridge: Cambridge University Press, 1996], 200).

101. H. Roche et al., "Early Hominid Stone Tool Production and Technical Skill 2.34 Myr Ago in West Turkana, Kenya," *Nature* 399 (1999): 57–60. See also Pelegrin, "Cognition and the Emergence of Language," 98–100.

102. PET scans done of human subjects as they flake stone indicate activation of brain areas associated with both motor activities and multimodal sensory inputs: "vision, touch, and proprioception, or sense of body position and motion" (Dietrich Stout et al., "Stone Tool-Making and Brain Activation: Positron Emission Tomography [PET] Studies," *Journal of Archaeological Science* 27 [2000]: 1215). See also Dietrich Stout and Thierry Chaminde, "The Evolutionary Neuroscience of Tool Making," *Neuropsychologia* 45 (2007): 1091–1100. For a discussion of the potential importance of "talking to oneself" to the evolution of consciousness and thinking, see Daniel C. Dennett, *Consciousness Explained* (Boston: Little, Brown, 1991), 193–226.

103. Gowlett, "Elements of Design Form in Acheulian Bifaces."

104. Schick and Toth, *Making Silent Stones Speak*, 258–260; Keeley, "Microwear Analysis of Lithics," 135–137.

105. William H. Calvin, "The Unitary Hypothesis: A Common Neural Circuitry for Novel Manipulations, Language, Plan-Ahead, and Throwing?" in *Tools, Language, and Cognition in Human Evolution*, ed. Kathleen R. Gibson and Tim Ingold (Cambridge: Cambridge University Press, 1993), 230–250.

106. Alan Walker and Pat Shipman, *The Wisdom of the Bones: In Search of Human Origins* (New York: Knopf, 1996).

107. Robert L. Kelly, *The Foraging Spectrum: Diversity in Hunter-Gatherer Lifeways* (Washington, D.C.: Smithsonian Institution Press, 1995).

108. Christoph. M. Monahan, "New Zooarchaeological Data from Bed II, Olduvai Gorge, Tanzania: Implications for Hominid Behavior in the Early Pleistocene," *Journal of Human Evolution* 31 (1996): 93–128.

109. See, for example, Arthur J. Jelinek, "The Lower Paleolithic: Current Evidence and Interpretation," *Annual Review of Anthropology* 6 (1977): 11–32; and Pitts and Roberts, *Fairweather Eden*, 287.

110. David Lordkipanidze et al., "Postcranial Evidence from Early *Homo* from Dmanisi, Georgia," *Nature* 449 (2007): 305–310.

111. For example, Grahame Clark, *World Prehistory in New Perspective* (Cambridge: Cambridge University Press, 1977), 30.

112. John R. Searle, *The Mind: A Brief Introduction* (New York: Oxford University Press, 2004), 19.

113. The late J. Desmond Clark is quoted as suggesting that the communications of the biface-makers probably resembled the bifaces: "the same things over and over and over again" (Schick and Toth, *Making Silent Stones Speak*, 280).

114. See, for example, John Wymer, *Lower Palaeolithic Archaeology in Britain as Represented by the Thames Valley* (London: Baker, 1968), 46–60.

115. François Bordes, *Typologie du Paléolithique ancient et moyen* (Bordeaux: Imprimeries Delmas, 1961); Derek Roe, "British Lower and Middle Palaeolithic Handaxe Groups," *Proceedings of the Prehistoric Society* 34 (1968): 245–267; P. Callow, "The Lower and Middle Palaeolithic of Britain and Adjacent Areas of Europe" (Ph.D. diss., Cambridge University, 1976).

116. Glynn Ll. Isaac, *Olorgesailie: Archeological Studies of a Middle Pleistocene Lake Basin in Kenya* (Chicago: University of Chicago Press, 1977), 116–145.

117. Shannon P. McPherron, "What Typology Can Tell Us About Acheulian Handaxe Production," in *Axe Age*, ed. Goren-Inbar and Sharon, 267–285.

118. Ofer Bar-Yosef and Anna Belfer-Cohen, "From Africa to Eurasia—Early Dispersals," *Quaternary International* 75 (2001): 19–28. The earliest dated occurrence of bifaces in North Africa seems to be at Casablanca (Morocco), where early Acheulean artifacts of quartzite and flint date to about a million years ago. See Jean-Paul Raynal et al., "The Earliest Occupation of North-Africa: The Moroccan Perspective," *Quaternary International* 75 (2001): 68–69.

119. G. Philip Rightmire, "Human Evolution in the Middle Pleistocene: The Role of *Homo heidelbergensis*," *Evolutionary Anthropology* 6 (1998): 218–227.

120. For example, Klein, *Human Career*, 330–343.

121. Bar-Yosef and Belfer-Cohen, "From Africa to Eurasia," 24.

122. Manuel Santonja and Paola Villa, "The Acheulean of Western Europe," in *Axe Age*, ed. Goren-Inbar and Sharon, 429–478. Bifaces are reported from Notarchirico (southern Italy) in deposits that underlie a volcanic tephra dating to 640,000 years ago.

123. Michael D. Petraglia, "The Lower Palaeolithic of India and Its Bearing on the Asian Record," in *Early Human Behaviour in Global Context: The Rise and Diversity of the Lower Paleolithic Record*, ed. Michael D. Petraglia and Ravi Korisettar (London: Routledge, 1998), 343–390.

124. Thomas Wynn and Forrest Tierson, "Regional Comparison of the Shapes of Later Acheulean Handaxes," *American Anthropologist* 92 (1990): 73–84.

3. Modern Humans and the Super-Brain

1.   Richard G. Klein, *The Human Career: Human Biological and Cultural Origins*, 3rd ed. (Chicago: University of Chicago Press, 2009).

2.   Francesco d'Errico et al., "Archaeological Evidence for the Emergence of Language, Symbolism, and Music—An Alternative Multidisciplinary Perspective," *Journal of World Prehistory* 17 (2003): 1–70.

3.   John F. Hoffecker, "Innovation and Technological Knowledge in the Upper Paleolithic of Northern Eurasia," *Evolutionary Anthropology* 14 (2005): 186–198.

4.   Christopher S. Henshilwood and Curtis W. Marean, "The Origin of Modern Behavior: Critique of the Models and Their Test Implications," *Current Anthropology* 44 (2003): 627–651.

5.   Lyn Wadley, "What Is Cultural Modernity? A General View and a South African Perspective from Rose Cottage Cave," *Cambridge Archaeological Journal* 1 (2001): 201–221. The concept of material "storage of symbols" was introduced in Merlin

Donald, *Origins of the Modern Mind: Three Stages in the Evolution of Culture and Cognition* (Cambridge, Mass.: Harvard University Press, 1991).

6. Noam Chomsky, *On Nature and Language* (Cambridge: Cambridge University Press, 2002), 45–46; Michael Corballis, "Recursion as the Key to the Human Mind," in *From Mating to Mentality: Evaluating Evolutionary Psychology*, ed. Kim Sterelny and Julie Fitness (New York: Psychology Press, 2003), 155–171.

7. According to Steven Pinker, "As far as biological cause and effect are concerned, music is useless. It shows no signs of design for attaining a goal such as long life, grandchildren, or accurate perception and prediction of the world" (*How the Mind Works* [New York: Norton, 1997], 528).

8. Claude Lévi-Strauss, *The Raw and the Cooked: Introduction to a Science of Mythology*, trans. John Weightman and Doreen Weightman (New York: Harper & Row, 1966), 1:18–19. Lévi-Strauss noted that while painters draw on natural color, "there are no musical sounds in nature."

9. d'Errico et al., "Archaeological Evidence for the Emergence of Language, Symbolism, and Music."

10. See, for example, William Noble and Iain Davidson, "The Evolutionary Emergence of Modern Human Behaviour: Language and Its Archaeology," *Man* 26 (1991): 223–253; and Paul Mellars, *The Neanderthal Legacy: An Archaeological Perspective from Western Europe* (Princeton, N.J.: Princeton University Press, 1996), 387–419.

11. Noam Chomsky, *The Minimalist Program* (Cambridge, Mass.: MIT Press), and *On Nature and Language*, 92–161.

12. Klein, *Human Career*, 514–517; Frederick L. Coolidge and Thomas Wynn, "Working Memory, Its Executive Functions, and the Emergence of Modern Thinking," *Cambridge Archaeological Journal* 15 (2005): 5–26.

13. Stanley H. Ambrose, "Late Pleistocene Human Population Bottlenecks, Volcanic Winter, and Differentiation of Modern Humans," *Journal of Human Evolution* 34 (1998): 623–651.

14. John J. Shea, "Neandertals, Competition, and the Origin of Modern Human Behavior in the Levant," *Evolutionary Anthropology* 12 (2003): 173–187. A recent report on the draft sequence of the Neanderthal genome based on ancient DNA suggests that genic exchange between Neanderthals and modern humans may have occurred in the Near East before 100,000 years ago. See Richard E. Green et al., "A Draft Sequence of the Neandertal Genome," *Science* 328 (2010): 710–722.

15. One aspect of modernity that may have evolved originally for a different function is that the neural structures underlying creative language use could have been derived from the evolution of throwing objects. See William H. Calvin, "The Unitary Hypothesis: A Common Neural Circuitry for Novel Manipulations, Language, Plan-Ahead, and Throwing?" in *Tools, Language, and Cognition in Human Evolution*, ed. Kathleen R. Gibson and Tim Ingold (Cambridge: Cambridge University Press, 1993), 230–250. See also, for example, Ian Tattersall, "Human Origins: Out of Africa," *Proceedings of the National Academy of Sciences* 106 (2009): 16020.

16. Tattersall, "Human Origins," 16018.

17. G. Philip Rightmire, "Human Evolution in the Middle Pleistocene: The Role of *Homo heidelbergensis*," *Evolutionary Anthropology* 6 (1998): 218–227.

18. On the energy and nutritional requirements of large brains, see T. Westermarck and E. Antila, "Diet in Relation to the Nervous System," in *Human Nutrition and Dietetics*, 10th ed., ed. J. S. Garrow, W. P. T. James, and A. Ralph (New York: Churchill Livingstone, 2000), 715–730. On the choking risk created by the modern vocal tract, see Philip Lieberman, *The Biology and Evolution of Language* (Cambridge, Mass.: Harvard University Press, 1984), 271–283.

19. Gerald M. Edelman and Giulio Tononi, *A Universe of Consciousness: How Matter Becomes Imagination* (New York: Basic Books, 2000), 37–50.

20. Gerhardt von Bonin, *The Evolution of the Human Brain* (Chicago: University of Chicago Press, 1963), 41–43; Philip V. Tobias, *The Brain in Hominid Evolution* (New York: Columbia University Press, 1971), 101–103.

21. Klein suggests that a mutation related to speech occurred roughly 50,000 years ago among anatomically modern humans in Africa ("neural hypothesis"), in *Human Career*, 647–653.

22. See, for example, G. Philip Rightmire, "*Homo* in the Middle Pleistocene: Hypodigms, Variation, and Species Recognition," *Evolutionary Anthropology* 17 (2008): 8–21.

23. Pamela R. Willoughby, *The Evolution of Modern Humans in Africa: A Comprehensive Guide* (Lanham, Md.: AltaMira Press, 2007), 17–194; Ralph L. Holloway et al., "Evolution of the Brain in Humans—Paleoneurology," in *Encyclopedia of Neuroscience*, ed. Mark D. Binder, Nobutaka Hirokawa, and Uwe Windhorst (New York: Springer, 2009), 1326–1334.

24. Rightmire, "*Homo* in the Middle Pleistocene," 12–13.

25. The pattern of cold-climate adaptation in Neanderthal anatomy was first noted in Carleton S. Coon, *The Origin of Races* (New York: Knopf, 1963), 542–548. More recent studies include Erik Trinkaus, "Neanderthal Limb Proportions and Cold Adaptation," in *Aspects of Human Evolution*, ed. Chris Stringer (London: Taylor & Francis, 1981), 187–224. For a current view, see Timothy D. Weaver, "The Meaning of Neandertal Skeletal Morphology," *Proceedings of the National Academy of Sciences* 106 (2009): 16028–16033. Weaver argues that genetic drift was a significant factor in many characteristic Neanderthal features.

26. Ralph L. Holloway, "The Poor Brain of *Homo sapiens neanderthalensis*: See What You Please . . . ," in *Ancestors: The Hard Evidence*, ed. Eric Delson (New York: Liss, 1985), 319–324.

27. Trenton W. Holliday, "Postcranial Evidence of Cold Adaptation in European Neandertals," *American Journal of Physical Anthropology* 104 (1997): 245–258. For a discussion of Neanderthal anatomy and technology, see John F. Hoffecker, *A Prehistory of the North: Human Settlement of the Higher Latitudes* (New Brunswick, N.J.: Rutgers University Press, 2005), 53–55.

28. Robin Dunbar, *Grooming, Gossip, and the Evolution of Language* (Cambridge, Mass.: Harvard University Press, 1996), and "The Social Brain Hypothesis," *Evolutionary Anthropology* 6 (1998): 178–190. The evolutionary biology of eusociality

may offer some insight into the selection pressures that lay behind language and the super-brain, despite the difference between the super-brain and a super-organism. True social behavior among the insects is thought to have evolved as a result of "progressive provisioning" of offspring by solitary females. In theory, selection would favor eusociality when the long-term reproductive fitness of offspring was enhanced by remaining at the nest site to help the parent provision larvae rather than by dispersing to reproduce as another solitary female. See Bert Hölldobler and E. O. Wilson, *The Superorganism: The Beauty, Elegance, and Strangeness of Insect Societies* (New York: Norton, 2009), 31-42. It has long been assumed that among early humans, provisioning of females and offspring became an increasingly important factor in long-term reproductive success as the demands of child care expanded.

29. Richard Byrne and Andrew Whiten, eds., *Machiavellian Intelligence: Social Expertise and the Evolution of Intellect in Monkeys, Apes, and Humans* (Oxford: Oxford University Press, 1988). See also Dunbar, *Grooming, Gossip, and the Evolution of Language*, 60-64.

30. Miriam Noël Haidle, "How to Think a Spear," in *Cognitive Archaeology and Human Evolution*, ed. Sophie A. de Beaune, Frederick L. Coolidge, and Thomas Wynn (Cambridge: Cambridge university Press, 2009), 57-73.

31. See, for example, Claude Lévi-Strauss, *The Savage Mind* (Chicago: University of Chicago Press, 1966).

32. Joseph LeDoux, *Synaptic Self: How Our Brains Become Who We Are* (New York: Penguin, 2002).

33. Perhaps spoken language and symbolism—and the super-brain they enabled—were at least in part a response to information overload in anatomically modern humans roughly 100,000 years ago. Language obviously provided a means of sharing information among individual brains, but it also would have enhanced the organization and storage of information. In Africa, modern humans would have been exposed to greater potential for information overload than would Neanderthals because the former occupied a more complex environmental setting than the latter. Brain size in modern humans subsequently declined, which would seem to reflect the decreased need for information storage in individual brains as greater quantities of information came to be stored outside the brain (and perhaps better organized). See, for example, Tobias, *Brain in Hominid Evolution*, 100-103.

34. Ralph L. Holloway, "Brief Communication: How Much Larger Is the Relative Volume of Area 10 of the Prefrontal Cortex in Humans?" *American Journal of Physical Anthropology* 118 (2002): 399-401; Daniel E. Lieberman, "Speculations About the Selective Basis for Modern Human Craniofacial Form," *Evolutionary Anthropology* 17 (2008): 55-68.

35. Marcus E. Raichle et al., "Practice-related Changes in Human Brain Functional Anatomy During Nonmotor Learning," *Cerebral Cortex* 4 (1994): 8-26; Peter Hagoort et al., "Integration of Word Meaning and World Knowledge in Language Comprehension," *Science* 304 (2004): 438-441; Elkhonon Goldberg, *The New Executive Brain: Frontal Lobes in a Complex World* (Oxford: Oxford University Press, 2009), 22-24.

36. Osbjorn M. Pearson, "Statistical and Biological Definitions of 'Anatomically Modern' Humans: Suggestions for a Unified Approach to Modern Morphology," *Evolutionary Anthropology* 17 (2008): 38–48. Chet S. Sherwood and colleagues also stress the likely importance of increases in the size of the cerebellum relative to body size in modern humans, as well as the connections between the cerebellum and the neocortex, observing that "these connections may be the anatomical substrate supporting the postulated cerebellar involvement in cognition, beyond its traditionally recognized role in motor coordination" ("Evolution of the Brain in Humans—Comparative Perspective," in *Encyclopedia of Neuroscience*, ed. Binder, Hirokawa, and Windhorst, 1336).

37. Jeffrey Laitman, "The Anatomy of Human Speech," *Natural History*, August 1984, 20–27; Lieberman, *Biology and Evolution of Language*, 271–283.

38. Leslie Aiello and Christopher Dean, *An Introduction to Human Evolutionary Anatomy* (London: Academic Press, 1990), 232–243.

39. Russell H. Tuttle, *Apes of the World: Their Social Behavior, Communication, Mentality and Ecology* (Park Ridge, N.J.: Noyes, 1986), 36–39.

40. Aiello and Dean, *Introduction to Human Evolutionary Anatomy*, 232.

41. See, for example, Ian Tattersall, *The Fossil Trail: How We Know What We Think We Know About Human Evolution* (New York: Oxford University Press, 1995), 211–212. Noam Chomsky has written about the difficulty of appreciating the strangeness of a phenomenon that is so familiar to all, in *Language and Mind*, 3rd ed. (Cambridge: Cambridge University Press, 2006), 21–23.

42. Chomsky wrote, "The use of language for communication might turn out to be a kind of epiphenomenon" (*On Nature and Language*, 107). Terrence W. Deacon suggested that language was similar to a virus that had co-evolved symbiotically with humans, in his widely read book *The Symbolic Species: The Co-Evolution of Language and the Brain* (New York: Norton, 1997). A similar argument was made earlier in Lewis Thomas, *The Lives of a Cell: Notes of a Biology Watcher* (New York: Viking Press, 1974). The thesis has been criticized rather sharply by Chomsky, who complained that it "seem[s] only to reshape standard problems of science as utter mysteries" (*On Nature and Language*, 83). In a more recent paper, Morten H. Christiansen and Nick Chater again advanced the notion that language represents an "interdependent organism" ("Language as Shaped by the Brain," *Brain and Behavioral Sciences* 32 [2008]: 489–558). Although I am skeptical of this idea (it seems to be a form of animism), the central theme of the paper makes sense to me; I also would guess that language has been "shaped by the brain" because it encodes internal mental representations for transmission in external form from one brain to another.

43. Marc D. Hauser, Noam Chomsky, and W. Tecumseh Fitch, "The Faculty of Language: What Is It, Who Has It, and How Did It Evolve?" *Science* 298 (2002): 1569–1579.

44. Lieberman, *Biology and Evolution of Language*, 287–313; Richard F. Kay, Matt Cartmill, and Michelle Balow, "The Hypoglossal Canal and the Origin of Human Vocal Behavior," *Proceedings of the National Academy of Sciences* 95 (1988): 5417–5419; Philip Lieberman, *Eve Spoke: Human Language and Human Evolution* (New

York: Norton, 1998); David De Gusta, W. Henry Gilbert, and Scott P. Turner, "Hypoglossal Canal Size and Hominid Speech," *Proceedings of the National Academy of Sciences* 96 (1999): 1800–1804; W. Tecumseh Fitch, "The Evolution of Speech: A Comparative Review," *Trends in Cognitive Sciences* 4 (2000): 258–267.

45. d'Errico et al., "Archaeological Evidence for the Emergence of Language, Symbolism, and Music," 27–31.

46. Green et al., "Draft Sequence of the Neandertal Genome"; Wolfgang Enard et al., "Molecular Evolution of FOXP2, a Gene Involved in Speech and Language," *Nature* 418 (2002): 869–872.

47. Stanley H. Ambrose, "Paleolithic Technology and Human Evolution," *Science* 291 (2001): 1748–1753.

48. James Deetz, *Invitation to Archaeology* (Garden City, N.Y.: Natural History Press, 1967), 83–101. Deetz subsequently described spoken language as material culture "in its gaseous state" (*In Small Things Forgotten: The Archaeology of American Life* [Garden City, N.Y.: Anchor Books, 1977], 24–25). The importance of multiple hierarchical levels for infinite creative potential is addressed in W. Tecumseh Fitch and Marc D. Hauser, "Computational Constraints on Syntactic Processing in a Nonhuman Primate," *Science* 303 (2004): 377–380.

49. For a discussion of the "generative interplay between the mental and material activities" of individuals and artifacts in the context of Europe 250,000 years ago, see Nathan Schlanger, "Understanding Levallois: Lithic Technology and Cognitive Archaeology," *Cambridge Archaeological Journal* 6 (1996): 231–254. This recalls the observations on refitted cores from a 2.3-million-year-old Oldowan site in East Africa described in chapter 2 (H. Roche et al., "Early Hominid Stone Tool Production and Technical Skill 2.34 Myr Ago in West Turkana, Kenya," *Nature* 399 [1999]: 57–60), although it is apparent that the number of hierarchically organized steps is significantly greater by 250,000 years ago.

50. Derek Bickerton argues that syntactic language was logically preceded by a nonsyntactic form of language comparable to that of very young children, in *Language and Species* (Chicago: University of Chicago Press, 1990), 130–163.

51. Eric Boëda, "Levallois: A Volumetric Construction, Methods, a Technique," in *The Definition and Interpretation of Levallois Technology*, ed. Harold L. Dibble and Ofer Bar-Yosef (Madison, Wis.: Prehistory Press, 1995), 41–68. See also Philip Van Peer, *The Levallois Reduction Strategy* (Madison, Wis.: Prehistory Press, 1992); and Mellars, *Neanderthal Legacy*. In recent decades, the analysis of prepared cores has shifted from an emphasis on finished core types to study of manufacturing processes.

52. Boëda, "Levallois," 46. See also Schlanger, "Understanding Levallois."

53. Boëda, "Levallois," 53.

54. Nicholas Rolland, "Levallois Technique Emergence: Single or Multiple? A Review of the Euro-African Record," in *Definition and Interpretation of Levallois Technology*, ed. Dibble and Bar-Yosef, 333–359. The emergence of prepared-core techniques in Africa has been tied for many years to the earlier "Victoria West" cores, but a direct link recently has been challenged in Stephen J. Lycett, "Are Victoria West Cores 'Proto-Levallois'? A Phylogenetic Assessment," *Journal of Human Evolution* 56 (2009): 175–191.

55. Sally McBrearty and Alison S. Brooks, "The Revolution That Wasn't: A New Interpretation of the Origin of Modern Human Behavior, " *Journal of Human Evolution* 39 (2000): 487–490. Lower Paleolithic artifacts from Peninj (Tanzania) dating to 1.6 to 1.4 million years ago are thought to reflect early experimentation with controlling flake products from cores. See Ignacio del Torre et al., "The Oldowan Industry of Peninj and Its Bearing on the Reconstruction of the Technological Skills of Lower Pleistocene Hominids," *Journal of Human Evolution* 44 (2003): 203-224.

56. See, for example, Robert Foley and Marta Mirazon Lahr, "On Stony Ground: Lithic Technology, Human Evolution, and the Emergence of Culture," *Evolutionary Anthropology* 12 (2003): 109–122.

57. Rolland, "Levallois Technique Emergence," 345–351; Mark White and Nick Ashton, "Lower Palaeolithic Core Technology and the Origins of the Levallois Method in North-Western Europe," *Current Anthropology* 44 (2003): 598–609.

58. Mellars, *Neanderthal Legacy*, 61–84.

59. For an overview of technology, including composite implements, among foraging peoples, see Wendell H. Oswalt, *An Anthropological Analysis of Food-Getting Technology* (New York: Wiley, 1976). Oswalt broke down individual implements into their component parts ("techno-units") for a comparative analysis of technological complexity.

60. Composite weapons in the form of microblade-inset spear points are known from several late Upper Paleolithic sites in northeastern Asia. See Craig M. Lee, "Microblade Technology in Beringia" [box], in John F. Hoffecker and Scott A. Elias, *Human Ecology of Beringia* (New York: Columbia University Press, 2007), 122-126.

61. Sylvia Beyries, "Functional Variability of Lithic Sets in the Middle Paleolithic," in *Upper Pleistocene Prehistory of Western Eurasia*, ed. Harold L. Dibble and Anta Montet-White (Philadelphia: University of Pennsylvania Museum, 1988), 213–224; Patricia Anderson-Gerfaud, "Aspects of Behavior in the Middle Palaeolithic: Functional Analysis of Stone Tools from Southwest France," in *The Emergence of Modern Humans: An Archaeological Perspective*, ed. Paul Mellars (Edinburgh: Edinburgh University Press, 1990), 389–418; Marlize Lombard, "Evidence of Hunting and Hafting During the Middle Stone Age at Sibudu Cave, KwaZulu-Natal: A Multianalytical Approach," *Journal of Human Evolution* 48 (2005): 279–300.

62. Eric Boëda et al., "Bitumen as a Hafting Material on Middle Palaeolithic Artefacts," *Nature* 380 (1996): 336-338; McBrearty and Brooks, "Revolution That Wasn't," 497; Bruce L. Hardy et al., "Stone Tool Function at the Paleolithic Sites of Starosele and Buran Kaya III, Crimea: Behavioral Implications," *Proceedings of the National Academy of Sciences* 98 (2001): 10972-10977; Marlize Lombard, "The Gripping Nature of Ochre: The Association of Ochre with Howiesons Poort Adhesives and Later Stone Age Mastics from South Africa," *Journal of Human Evolution* 53 (2007): 406-419.

63. Paola Villa et al., "The Howiesons Poort and MSA III at Klasies River Main Site, Cave 1A," *Journal of Archaeological Science* 37 (2010): 630-655.

64. Archaeologists interested in the development of human cognitive faculties sometimes have been unimpressed with Levallois prepared-core technology. William

Noble and Iain Davidson were skeptical of the intentional control over blank size and shape and, more generally, found the technique "not important" (*Human Evolution, Language and Mind: A Psychological and Archaeological Inquiry* [Cambridge: Cambridge University Press, 1996], 200–201). Thomas Wynn thought the required concepts of Levallois core technology "no more complex than those required for fine bifaces" ("The Evolution of Tools and Symbolic Behaviour," in *Handbook of Human Symbolic Evolution*, ed. Andrew Lock and Charles R. Peters [Oxford: Blackwell, 1999], 273). These views do not seem to take into account the likely connection between Levallois cores and composite implements.

65. Ambrose, "Paleolithic Technology and Human Evolution," 1751.

66. James B. Rowe et al., "The Prefrontal Cortex: Response Selection or Maintenance Within Working Memory?" *Science* 288 (2000): 1656–1560; Coolidge and Wynn, "Working Memory." See also Cordelia Fine, ed., *The Britannica Guide to the Brain* (London: Constable and Robinson, 2008), 112–118; and Goldberg, *New Executive Brain*, 92–98.

67. Hagoort et al., "Integration of Word Meaning and World Knowledge in Language Comprehension"; Goldberg, *New Executive Brain*, 23–24.

68. John F Hoffecker, "Representation and Recursion in the Archaeological Record," *Journal of Archaeological Method and Theory* 14 (2007): 359–387.

69. Katherine Scott, "Two Hunting Episodes of Middle Palaeolithic Age at La Cotte de Saint-Brelade, Jersey (Channel Islands)," *World Archaeology* 12 (1980): 137–152; Hervé Bocherens et al., "Isotopic Evidence for Diet and Subsistence Pattern of the Saint Cesaire I Neanderthal: Review and Use of a Multi-source Mixing Model," *Journal of Human Evolution* 49 (2005): 71–87.

70. McBrearty and Brooks, "Revolution That Wasn't," 510–513; Klein, *Human Career*, 555–564.

71. Mellars, *Neanderthal Legacy*, 369–371; Ian Watts, "The Origin of Symbolic Culture," in *The Evolution of Culture: An Interdisciplinary View*, ed. Robin Dunbar, Chris Knight, and Camilla Power (New Brunswick, N.J.: Rutgers University Press, 1999), 113–146.

72. See, for example, John J. Shea, "Middle Paleolithic Spear Point Technology," in *Projectile Technology*, ed. Heidi Knecht (New York: Plenum Press, 1997), 79–106.

73. Lombard, "Gripping Nature of Ochre," 406–419.

74. Lyn Wadley, "Revisiting Cultural Modernity and the Role of Red Ochre in the Middle Stone Age," in *The Prehistory of Africa: Tracing the Lineage of Modern Man*, ed. Himla Soodyall (Johannesburg: Jonathan Ball, 2006), 49–63.

75. J. Desmond Clark, "African and Asian Perspectives on the Origin of Modern Humans," in *The Origin of Modern Humans and the Impact of Chronometric Dating*, ed. M. J. Aitken, C. B. Stringer, and P. A. Mellars (Princeton, N.J.: Princeton University Press, 1993), 148–178.

76. François Bordes, "Mousterian Cultures in France," *Science* 134 (1961): 803–810.

77. Lewis R. Binford and Sally R. Binford, "A Preliminary Analysis of Functional Variability in the Mousterian of Levallois Facies," *American Anthropologist* 68 (1966): 238–295.

78. Ongoing research on composite implements may eventually reveal greater complexity in the African technology, providing an additional contrast (other than the apparent lack of regional differentiation in artifact style) with that of the Neanderthals. For example, the alternative hafting locations (that is, end versus side) from southern Africa (recently reported in Villa et al., "Howiesons Poort and MSA III at Klasies River Main Site, Cave 1A") may not be present in Eurasia.

79. See, for example, Francesco d'Errico, "The Invisible Frontier: A Multiple Species Model for the Origin of Behavioral Modernity," *Evolutionary Anthropology* 12 (2003): 188–202.

80. A prominent critic of the perceived parallels between technology and language is Thomas Wynn, who observed that "the action sequences of tool-making and tool-use are organized and learned in a simple way. One action is tied to the preceding action, and long strings of such connections are learned by rote and repetition. This is very unlike the production of a sentence" ("Tools, Grammar and the Archaeology of Cognition," *Cambridge Archaeological Journal* 1 [1991]: 198). See also Thomas Wynn, "Handaxe Enigmas," *World Archaeology* 27 (1995): 10–24. Wynn's view may reflect an emphasis on simple technologies; I find it difficult to avoid parallels between language structure and the complex hierarchical organization of more recent machines and devices.

81. Coolidge and Wynn, "Working Memory."

82. Karen Emmorey, Sonya Mehta, and Thomas J. Grabowski, "The Neural Correlates of Sign Versus Word Production," *NeuroImage* 36 (2007): 202–208.

83. Humans who lived 300,000 to 250,000 years ago (and perhaps much earlier) probably communicated a substantial amount of information with a wide range of sounds. Vocal communication among vervet monkeys provides a possible model. In a classic study, *How Monkeys See the World: Inside the Mind of Another Species* (Chicago: University of Chicago Press, 1990), Dorothy L. Cheney and Robert S. Seyfarth reported that these highly social monkeys identify various predators with specific sounds (that is, vocal iconic symbols). Monkeys and apes also communicate with a variety of gestures, primarily using the hands and feet and sometimes employing objects. See Josep Call and Michael Tomasello, *The Gestural Communication of Apes and Monkeys* (Mahwah, N.J.: Erlbaum, 2007).

84. Lieberman, *Biology and Evolution of Language.*

85. See, for example, Noble and Davidson, "Evolutionary Emergence of Modern Human Behaviour." In a provocative paper, "The Archaeology of Perception: Traces of Depiction and Language," *Current Anthropology* 30 (1989): 125–155, Iain Davidson and William Noble suggested that visual art was an essential stepping stone to language—rather than merely a byproduct or consequence of it. This thesis received little support from other archaeologists—most of whom seem to believe that language either preceded or accompanied the appearance of visual art—but it offers a parallel to many of the arguments presented here (for example, that interaction and feedback between brain and artifact—between internal and external mental representation—was a catalyst for the development of human cognitive faculties).

86. Dietrich Mania and Ursula Mania, "Deliberate Engravings on Bone Artefacts of *Homo erectus*," *Rock Art Research* 5 (1988): 91–107; Klein, *Human Career*, 407–410.

87. Francesco d'Errico and April Nowell, "A New Look at the Berekhat Ram Figurine: Implications for the Origins of Symbolism," *Cambridge Archaeological Journal* 10 (2000): 123-167.

88. Marian Vanhaeren et al., "Middle Paleolithic Shell Beads in Algeria and Israel," *Science* 312 (2006): 1785-1788. Indirect evidence for the use of plant and/or animal materials for line extends back to the indications of hafting discussed earlier (that is, around 250,000 years ago).

89. Wadley, "What Is Cultural Modernity," 208; Marian Vanhaeren and Francesco d'Errico, "Aurignacian Ethno-linguistic Geography of Europe Revealed by Personal Ornaments," *Journal of Archaeological Science* 33 (2006): 1105-1128.

90. Mellars, *Neanderthal Legacy*, 374-375; d'Errico, "Invisible Frontier," 198-199. Personal ornaments are associated with Neanderthal skeletal remains in several occupation layers in southwestern Europe assigned to the "Chatelperronian" industry, but these may be cases where Middle Paleolithic artifacts and skeletal remains were mixed with overlying Upper Paleolithic artifacts by frost action. See, for example, Klein, *Human Career*, 748-750.

91. Christopher Henshilwood, "Fully Symbolic *Sapiens* Behaviour: Innovation in the Middle Stone Age at Blombos Cave, South Africa," in *Rethinking the Human Revolution: New Behavioural and Biological Perspectives on the Origin and Dispersal of Modern Humans*, ed. Paul Mellars et al. (Cambridge: McDonald Institute for Archaeological Research, 2007), 123-132.

92. Clive Gamble, *Timewalkers: The Prehistory of Global Colonization* (Cambridge, Mass.: Harvard University Press, 1994), 181-202; Hoffecker, *Prehistory of the North*, 75-91.

93. Ofer Bar-Yosef, "The Middle and Early Upper Paleolithic in Southwest Asia and Neighboring Regions," in *The Geography of Neandertals and Modern Humans in Europe and the Greater Mediterranean*, ed. Ofer Bar-Yosef and David Pilbeam (Cambridge, Mass.: Peabody Museum of Archaeology and Ethnography, 2000), 107-156.

94. Hoffecker, "Innovation and Technological Knowledge in the Upper Paleolithic," 187-190.

95. Hoffecker, "Representation and Recursion in the Archaeological Record," 377-378.

96. d'Errico et al., "Archaeological Evidence for the Emergence of Language, Symbolism, and Music," 39-48.

97. Hoffecker, "Innovation and Technological Knowledge in the Upper Paleolithic," 190-196.

4. The Upper Paleolithic as History

1. V. Gordon Childe, "Retrospect," *Antiquity* 32 (1958): 70. See also V. Gordon Childe, *The Dawn of European Civilization* (London: Kegan Paul, 1925). Bruce G. Trigger notes that this book had been preceded by several syntheses of European prehistory in Britain, but that it was the "original and convincing manner in which Childe analysed and presented his data" that accounted for its impact (*Gordon Childe: Revolutions in Archaeology* [New York: Columbia University Press,

1980], 37–40). See also Bruce G. Trigger, "Childe's Relevance to the 1990s," in *The Archaeology of V. Gordon Childe*, ed. David R. Harris (Chicago: University of Chicago Press, 1994), 9–34. The essay begins with the observation that Childe, "although dead since 1957, remains the most renowned and widely read archaeologist of the 20th century."

2.   Trigger, *Gordon Childe*, 91.

3.   See, for example, V. Gordon Childe, "Races, Peoples, and Cultures in Prehistoric Europe," *History* 18 (1933): 193–203.

4.   V. Gordon Childe wrote that "food production . . . was an economic revolution. . . . It opened up a richer and more reliable supply of food. . . . Judging by the observed effects of the Industrial Revolution in England, a rapid increase of population would be the normal corollary of such a change" (*New Light on the Most Ancient East: The Oriental Prelude to European Prehistory* [London: Kegan Paul, 1934], 42). And regarding the rise of urban centers, he wrote "in response to the opportunities of livelihood created by the new economy an industrial proletariat multiplied as quickly as it did in England during the industrial revolution" (186). The concept of the "Industrial Revolution" was articulated in a series of lectures delivered at Oxford University in 1880 and 1881 by the elder Arnold Toynbee and later recalled and published by his students. See Arnold Toynbee, *The Industrial Revolution* (Boston: Beacon Press, 1956). See also Kevin Greene, "V. Gordon Childe and the Vocabulary of Revolutionary Change," *Antiquity* 73 (1999): 97–109.

5.   Clive Gamble, *Origins and Revolutions: Human Identity in Earliest Prehistory* (Cambridge: Cambridge University Press, 2007), 10–16.

6.   V. Gordon Childe, *Man Makes Himself* (London: Watts, 1936).

7.   V. Gordon Childe, *What Happened in History* (Harmondsworth: Penguin, 1942).

8.   Trigger, *Gordon Childe*, 17–19; V. Gordon Childe, *Scotland Before the Scots* (London: Methuen, 1946). In a personal recollection published in the early 1990s, Howard Kilbride-Jones wrote, "One must see everything in terms of Childe the exhibitionist. . . . I always had a feeling that Childe's Marxism was an umbrella under which he sheltered in order to be different from the rest, meaning of course the Establishment" ("From Mr Howard Kilbride-Jones, 6 April 1992," in *Archaeology of V. Gordon Childe*, ed. Harris, 135–139). Sally Green suggested the reverse—that Childe's tongue-in-cheek comments were an umbrella under which he sheltered a sincere commitment to Marxism—in *Prehistorian: A Biography of V. Gordon Childe* (Bradford-on-Avon: Moonraker Press, 1981), 77. But his later writings make it clear that he saw little value in the way in which Marxist philosophy had been applied to archaeology in the Soviet Union.

9.   Trigger, *Gordon Childe*, 124.

10.   Ibid., 131. See also Trigger, "Childe's Relevance to the 1990s," 23. It should be noted, however, that Childe made no reference to R. G. Collingwood in his article "Retrospect," which was published posthumously.

11.   R. G. Collingwood, *The Idea of History* (Oxford: Clarendon Press, 1946). This book was edited and published several years after Collingwood's death.

12.   Ibid., 217.

13. Trigger, *Gordon Childe*, 121.
14. V. Gordon Childe, *History* (London: Cobbett Press, 1947), 67–83. This book contains references to Marx, Engels, Lenin, and even Comrade Stalin.
15. V. Gordon Childe, *Social Evolution* (London: Watts, 1951), 1. In this book, Childe explicitly criticized the approach of Soviet archaeology—which had so impressed him in 1935—noting that it "assumes in advance precisely what archaeological facts have to prove." (29). And he later wrote, "[A]t last I rid my mind of transcendental laws determining history and mechanical causes, whether economic or environmental, automatically shaping its course" ("Retrospect," 73).
16. V. Gordon Childe, *Society and Knowledge: The Growth of Human Traditions* (New York: Harper, 1956). Although this book contains only one—somewhat critical—reference to the application of Marxist ideas to society and economics (102–103), Childe could not resist a comparison between "the Christian church under the Roman Empire [and] the Communist Party in the U.S.A." in a discussion of groups within larger societies (101–102).
17. Ibid., 79.
18. Ibid., 124.
19. Collingwood, *Idea of History*, 215.
20. Ibid., 209.
21. Childe, *Society and Knowledge*, 1.
22. V. Gordon Childe, *Piecing Together the Past: The Interpretation of Archaeological Data* (New York: Praeger, 1956), 1. Both Collingwood and Childe were swimming against the current. As Daniel Lord Smail observes, modern historians managed to sidestep Darwin and preserve the biblical time frame of the human past: "[T]he sacred was deftly translated into a secular key: the Garden of Eden became the irrigated fields of Mesopotamia, and the creation of man was reconfigured as the rise of civilization. Prehistory . . . was cantilevered outside the narrative buttresses that sustain the edifice of Western civilization" (*On Deep History and the Brain* [Berkeley: University of California Press, 2008], 4).
23. According to Childe, "[C]raft processes and products that archaeologists can observe can stand beside mathematical tablets and surgical papyri as genuine documents in the history of science" ("Retrospect," 72).
24. Bruce G. Trigger, *A History of Archaeological Thought* (Cambridge: Cambridge University Press, 1989), 94–102.
25. See, for example, Henry Fairfield Osborn, *Men of the Old Stone Age: Their Environment, Life, and Art* (New York: Scribner, 1915); and Denis Peyrony, *Élements de préhistoire* (Ussel: Jacques Eyboulet, 1934).
26. The Upper Paleolithic was identified with the development of matrilineal clan society and the preceding Middle Paleolithic with the "promiscuous horde" of the Neanderthals. See, for example, P. P. Efimenko, *Pervobytnoe obshchestvo*, 2nd ed. (Leningrad: USSR Academy of Sciences, 1938).
27. John F. Hoffecker, "Innovation and Technological Knowledge in the Upper Paleolithic of Northern Eurasia," *Evolutionary Anthropology* 14 (2005): 196.
28. Collingwood, *Idea of History*, 215.

29. These dates reflect the application of a calibration curve to the radiocarbon chronology of the Upper Paleolithic. Past fluctuations in atmospheric radiocarbon produce variable results in age estimates of carbon samples, and the cosmogenic radionuclide peak of 40,000 years ago corresponds to especially high concentrations of 14C, which yields age estimates that are several thousand years younger than the calendrical scale.

30. Hoffecker, "Innovation and Technological Knowledge in the Upper Paleolithic," 195–196.

31. John F. Hoffecker, "Representation and Recursion in the Archaeological Record," *Journal of Archaeological Method and Theory* 14 (2007): 359–387.

32. Francesco d'Errico et al., "Archaeological Evidence for the Emergence of Language, Symbolism, and Music—An Alternative Multidisciplinary Perspective," *Journal of World Prehistory* 17 (2003): 1–70.

33. The comparison between EUP visual art and music (instruments) and EUP technology can be drawn only for the late EUP, because there is little evidence for any visual art or music making before 40,000 years ago, and this may be due entirely to sampling and preservation bias.

34. For a discussion of how visual art may be related to ideas about space, time, and light in the postmedieval world, see Leonard Shlain's interesting book *Art and Physics: Parallel Visions in Space, Time, and Light* (New York: Morrow, 1991).

35. This is a complex issue long discussed by historians of technology as well as others. See, for example, Thorstein Veblen, *The Instinct of Workmanship and the State of the Industrial Arts* (New York: Macmillan, 1914); and Lynn White, *Medieval Religion and Technology: Collected Essays* (Berkeley: University of California Press, 1978).

36. See, for example, Philip Allsworth-Jones, "The Szeletian and the Stratigraphic Succession in Central Europe and Adjacent Areas: Main Trends, Recent Results, and Problems for Resolution," in *The Emergence of Modern Humans: An Archaeological Perspective*, ed. Paul Mellars (Edinburgh: Edinburgh University Press, 1990), 160–242.

37. John F. Hoffecker, "The Spread of Modern Humans in Europe," *Proceedings of the National Academy of Sciences* 106 (2009): 16040–16045.

38. Paul Mellars, "Going East: New Genetic and Archaeological Perspectives on the Modern Human Colonization of Eurasia," *Science* 313 (2006): 796–800. Although slightly dated now, the best general account of the global dispersal of modern humans remains Brian Fagan, *The Journey from Eden: The Peopling of Our World* (London: Thames and Hudson, 1990). For a more current overview, see Alice Roberts, *The Incredible Human Journey: The Story of How We Colonised the Planet* (London: Bloomsbury, 2009).

39. Rhys Jones, "East of Wallace's Line: Issues and Problems in the Colonisation of the Australian Continent," in *The Human Revolution: Behavioural and Biological Perspectives on the Origins of Modern Humans*, ed. Paul Mellars and Chris Stringer (Princeton, N.J.: Princeton University Press, 1989), 743–782; Philip J. Hapgood and Natalie R. Franklin, "The Revolution That Didn't Arrive: A Review of Pleistocene Sahul," *Journal of Human Evolution* 55 (2008): 187–222. Given the origins

of the Australian EUP (that is, watercraft), according to Hapgood and Franklin, the absence of earlier evidence for coastal marine resource use is likely due to rising sea levels, which probably inundated the earlier sites (203).

40. Hapgood and Franklin, "Revolution That Didn't Arrive," 192-201.

41. Ibid., 207-211. Bone implements at Bone Cave (Tasmania) were recovered in deposits that yielded dates as early 33,385 ± 554 calibrated radiocarbon years ago, but otherwise are unknown in the Australian EUP. It should be noted that bone implements are scarce in the oldest Upper Paleolithic assemblages of the Near East and Europe, despite their earlier presence in the African Middle Stone Age.

42. Anthony E. Marks and C. Reid Ferring, "The Early Upper Paleolithic of the Levant," in The Early Upper Paleolithic: Evidence from Europe and the Near East, ed. John F. Hoffecker and Cornelia A. Wolf (Oxford: British Archaeological Reports, 1988), 43-72; Ofer Bar-Yosef, "The Middle and Early Upper Paleolithic in Southwest Asia and Neighboring Regions," in The Geography of Neandertals and Modern Humans in Europe and the Greater Mediterranean, ed. Ofer Bar-Yosef and David Pilbeam (Cambridge, Mass.: Peabody Museum of Archaeology and Ethnology, 2000), 107-156. Bar-Yosef describes the transformation of stone industry in the Levant as a "technological revolution" (142-143).

43. Steven L. Kuhn et al., "The Early Upper Paleolithic Occupations at Üçağizli Cave (Hatay, Turkey)," Journal of Human Evolution 56 (2009): 87-113.

44. Miryam Bar-Mathews and Avner Ayalon, "Climatic Conditions in the Eastern Mediterranean During the Last Glacial (60-10 ky) and Their Relations to the Upper Palaeolithic in the Levant as Inferred from Oxygen and Carbon Isotope Systematics of Cave Deposits," in More Than Meets the Eye: Studies on Upper Palaeolithic Diversity in the Near East, ed. A. Nigel Goring-Morris and Anna Belfer-Cohen (Oxford: Oxbow Books, 2003), 13-18.

45. See, for example, William H. McNeill, Plagues and Peoples (Garden City, N.Y.: Anchor Books, 1976); and Brian Fagan, Floods, Famines, and Emperors: El Niño and the Fate of Civilizations (New York: Basic Books, 1999).

46. Ted Goebel, "The Pleistocene Colonization of Siberia and Peopling of the Americas: An Ecological Approach," Evolutionary Anthropology 8 (1999): 208-227.

47. Erik Trinkaus, "Neanderthal Limb Proportions and Cold Adaptation," in Aspects of Human Evolution, ed. Chris Stringer (London: Taylor & Francis, 1981), 187-224; Trenton W. Holliday, "Brachial and Crural Indices of European Late Upper Paleolithic and Mesolithic Humans," Journal of Human Evolution 36 (1999): 549-566.

48. John F. Hoffecker, A Prehistory of the North (New Brunswick, N.J.: Rutgers University Press, 2005), 75-82.

49. Hoffecker, "Spread of Modern Humans in Europe," 16044.

50. A new statistical approach has been applied to the classification of isolated dental remains from these sites, concluding that most of them can probably be assigned to modern humans. See Shara E. Bailey, Timothy D. Weaver, and Jean-Jacques Hublin, "Who Made the Aurignacian and Other Early Upper Paleolithic Industries?" Journal of Human Evolution 57 (2009): 11-26.

51. Steven L. Kuhn and Amilcare Bietti, "The Late Middle and Early Upper Paleolithic in Italy," in *Geography of Neandertals and Modern Humans in Europe and the Greater Mediterranean*, ed. Bar-Yosef and Pilbeam, 49–76.

52. Vance T. Holliday et al., "Geoarchaeology of the Kostenki–Borshchevo Sites, Don River Valley, Russia," *Geoarchaeology: An International Journal* 22 (2007): 181–228; John F. Hoffecker et al., "From the Bay of Naples to the River Don: The Campanian Ignimbrite Eruption and the Middle to Upper Paleolithic Transition in Eastern Europe," *Journal of Human Evolution* 55 (2008): 858–870.

53. Richard G. Klein, *Man and Culture in the Late Pleistocene: A Case Study* (San Francisco: Chandler, 1969), 29–31.

54. John F. Hoffecker et al., "Evidence for Kill-Butchery Events of Early Upper Paleolithic Age at Kostenki, Russia," *Journal of Archaeological Science* 37 (2010): 1073–1089.

55. Francesco G. Fedele, Biagio Giaccio, and Irka Hajdas, "Timescales and Cultural Processes at 40,000 BP in the Light of the Campanian Ignimbrite Eruption, Western Eurasia," *Journal of Human Evolution* 55 (2008): 834–857.

56. Paola Villa, François Bon, and Jean-Christophe Castel, "Fuel, Fire and Fireplaces in the Palaeolithic of Western Europe," *Review of Archaeology* 23 (2002): 33–42.

57. The earliest known eyed needle in eastern Europe dates to about 40,000 years ago and was recovered from Mezmaiskaya Cave (Russia) in the northwestern Caucasus. See Liubov V. Golovanova et al., "Significance of Ecological Factors in the Middle to Upper Paleolithic Transition," *Current Anthropology* 51 (2010): 655–691.

58. Heidi Knecht, "Splits and Wedges: The Techniques and Technology of Early Aurignacian Antler Working," in *Before Lascaux: The Complex Record of the Early Upper Paleolithic*, ed. Heidi Knecht, Ann Pike-Tay, and Randall White (Boca Raton, Fla.: CRC Press, 1993), 137–162.

59. Nicholas J. Conard, "Palaeolithic Ivory Sculptures from Southwestern Germany and the Origins of Figurative Art," *Nature* 426 (2003): 830–832, and "A Female Figurine from the Basal Aurignacian of Hohle Fels Cave in Southwestern Germany," *Nature* 459 (2009): 248–252.

60. d'Errico et al., "Archaeological Evidence for the Emergence of Language, Symbolism, and Music," 39–48.

61. This burial, which contained the poorly preserved remains of a child, was discovered by A. N. Rogachev at Kostenki 15 in 1952 and has since been dated by radiocarbon to about 30,000 calibrated years ago. Several years later, Rogachev excavated another burial in a later EUP level at Kostenki 14, but the dating of this skeleton has been problematic—it yielded several younger radiocarbon dates in recent years. See A. N. Rogachev, "Mnogosloinye Stoyanki Kostenkovsko-Borshevskogo raiona na Donu i Problema Razvitiya Kul'tury v Epokhy Verkhnego Paleolita na Russkoi Ravnine," *Materialy i Issledovaniya po Arkheologii SSSR* 59 (1957): 9–134; and Klein, *Man and Culture in the Late Pleistocene*, 94–96.

62. Ernest Becker, *The Denial of Death* (New York: Free Press, 1973).

63. The most famous example of grave offerings from a Neanderthal burial is the evidence of flowers, based on concentrations of preserved pollen, reported from Shanidar Cave (Iraq) by Ralph S. Solecki, *Shanidar, the First Flower People*

(New York: Knopf, 1971). It now appears likely, however, that the pollen concentrations are derived from flower heads brought into the cave by burrowing rodents. See Jeffrey D. Sommer, "The Shanidar IV 'Flower Burial': A Re-evaluation of Neanderthal Burial Ritual," *Cambridge Archaeological Journal* 9 (1999): 127–129.

64. Childe was not a Paleolithic specialist, but he did discuss early prehistory briefly in several books. His final comments on the Gravettian would seem to be in the posthumously published V. Gordon Childe, *The Prehistory of European Society* (Harmondsworth: Penguin, 1958).

65. Clive Gamble, *The Palaeolithic Settlement of Europe* (Cambridge: Cambridge University Press, 1986); Jiří Svoboda, Vojen Ložek, and Emanuel Vlček, *Hunters Between East and West: The Paleolithic of Moravia* (New York: Plenum Press, 1996), 131–170.

66. Goebel, "Pleistocene Colonization of Siberia and Peopling of the Americas," 216–218.

67. See, for example, Lawrence G. Straus, "The Upper Paleolithic of Europe: An Overview," *Evolutionary Anthropology* 4 (1995): 4–16.

68. Wendell H. Oswalt, *The Anthropological Analysis of Food-Getting Technology* (New York: Wiley, 1976).

69. Hoffecker, *Prehistory of the North*, 91–94.

70. See, for example, Childe, *Man Makes Himself*, 257–270. This is the final chapter of the book: "The Acceleration and Retardation of Progress."

71. See, for example, David S. Landes, *The Wealth and Poverty of Nations: Why Some Are So Rich and Some So Poor* (New York: Norton, 1999).

72. John F. Hoffecker, "The Eastern Gravettian 'Kostenki Culture' as an Arctic Adaptation," *Anthropological Papers of the University of Alaska*, n.s., 2 (2002): 115–136.

73. Michael P. Richards et al., "Stable Isotope Evidence for Increasing Dietary Breadth in the European Mid-Upper Paleolithic," *Proceedings of the National Academy of Sciences* 98 (2001): 6528–6532.

74. Olga Soffer et al., "Palaeolithic Perishables Made Permanent," *Antiquity* 74 (2000): 812–821; Zbigniew M. Bochenski et al., "Fowling During the Gravettian: The Avifauna of Pavlov I, Czech Republic," *Journal of Archaeological Science* 36 (2009): 2655–2665.

75. Pawel Valde-Nowak, Adam Nadachowski, and Mieczyslaw Wolsan, "Upper Palaeolithic Boomerang Made of a Mammoth Tusk in South Poland," *Nature* 329 (1987): 436–438.

76. Hoffecker, "Eastern Gravettian 'Kostenki Culture' as an Arctic Adaptation," 122–123.

77. Pamela P. Vandiver et al., "The Origins of Ceramic Technology at Dolni Věstonice, Czechoslovakia," *Science* 246 (1989): 1002–1008.

78. Hoffecker, "Eastern Gravettian 'Kostenki Culture' as an Arctic Adaptation," 120–123.

79. Roy A. Rappaport, *Ritual and Religion in the Making of Humanity* (Cambridge: Cambridge University Press, 1999).

80. Childe, *Man Makes Himself*, 261–270.

81. Clive Gamble, *The Palaeolithic Societies of Europe* (Cambridge: Cambridge University Press, 1999), 404–414.

82. Gamble, *Palaeolithic Settlement of Europe*, 324–331.

83. Pavel Dolukhanov, Dmitry Sokoloff, and Anvar Shukurov, "Radiocarbon Chronology of Upper Palaeolithic Sites in Eastern Europe at Improved Resolution," *Journal of Archaeological Science* 28 (2001): 699–712; Kelly E. Graf, "The Good, the Bad, and the Ugly: Evaluating the Radiocarbon Chronology of the Middle and Late Upper Paleolithic in the Enisei River Valley, South-Central Siberia," *Journal of Archaeological Science* 36 (2009): 694–707.

84. Hoffecker, *Prehistory of the North*, 94–95.

85. Hoffecker, "Innovation and Technological Knowledge in the Upper Paleolithic," 193–195.

86. Aaron Jonas Stutz, Natalie D. Munro, and Guy Bar-Oz, "Increasing the Resolution of the Broad Spectrum Revolution in the Southern Levantine Epipaleolithic (19–12 ka)," *Journal of Human Evolution* 56 (2009): 294–306.

87. According to Childe, "The bow . . . is perhaps the first engine man devised. The motive power is, indeed, just human muscular energy, but in the tension of the bow energy gradually expended in bending it is accumulated so as to be released all at once and concentrated in dispatching the arrow. The spear-thrower ingeniously augments the energy a man's arm can impart to a missile on the principle of the lever" (*Man Makes Himself*, 67).

88. Pierre Cattelain, "Un crochet de propulseur solutréen de la grotte de Combe-Saunière 1 (Dordogne)," *Bulletin de la Société préhistorique française* 86 (1989): 213–216.

89. Carl Mitcham, *Thinking Through Technology: The Path Between Engineering and Philosophy* (Chicago: University of Chicago Press, 1994), 20–24.

90. Grahame Clark, *Prehistoric Europe: The Economic Basis* (London: Methuen, 1952), 30–31.

91. I. G. Pidoplichko, *Mezhirichskie zhilishcha iz Kostei Mamonta* (Kiev: Naukova dumka, 1976), 164–167; Hoffecker, "Innovation and Technological Knowledge in the Upper Paleolithic," 194–195.

92. Thomas Hobbes, *Leviathan, or the Matter, Forme and Power of a Common-wealth Ecclesiasticall and Civill* (London: Penguin, 1968), 81.

93. Elisabetta Boaretto et al., "Radiocarbon Dating of Charcoal and Bone Collagen Associated with Early Pottery at Yuchanyan Cave, Hunan Province, China," *Proceedings of the National Academy of Sciences* 106 (2009): 9595–9600.

94. Charles T. Keally, Yasuhiro Taniguchi, and Yaroslav V. Kuzmin, "Understanding the Beginnings of Pottery Technology in Japan and Neighboring East Asia," *Review of Archaeology* 24 (2003): 3–14.

95. Ibid., 5.

96. New evidence for EUP dog domestication has been reported in Mietje Germonpré et al., "Fossil Dogs and Wolves from Palaeolithic Sites in Belgium, the Ukraine, and Russia: Osteometry, Ancient DNA, and Stable Isotopes," *Journal of Archaeological Science* 36 (2009): 473–490.

97. Norbert Benecke, "Studies on Early Dog Remains from Northern Europe," *Journal of Archaeological Science* 14 (1987): 31–49; Mikhail V. Sablin and Gennady A. Khlopachev, "The Earliest Ice Age Dogs: Evidence from Eliseevichi I," *Current Anthropology* 43 (2002): 795–799.

98. Peter Savolainen et al., "Genetic Evidence for an East Asian Origin for Domesticated Dogs," *Science* 298 (2002): 1610–1613.

99. Francesco d'Errico, "Palaeolithic Origins of Artificial Memory Systems: An Evolutionary Perspective," in *Cognition and Material Culture: The Archaeology of Symbolic Storage*, ed. Colin Renfrew and Chris Scarre (Cambridge: McDonald Institute for Archaeological Research, 1998), 19–50.

100. Alexander Marshack, *The Roots of Civilization: The Cognitive Beginnings of Man's First Art, Symbol and Notation* (New York: McGraw-Hill, 1972).

101. d'Errico, "Palaeolithic Origins of Artificial Memory Systems," 43.

102. I. G. Pidoplichko, *Pozdnepaleoliticheskie zhilishcha iz kostei mamonta na Ukraine* (Kiev: Naukova dumka, 1969), and *Mezhirichskie zhilishcha*; Zoya A. Abramova, "Two Examples of Terminal Paleolithic Adaptations," in *From Kostenki to Clovis: Upper Paleolithic–Paleo-Indian Adaptations*, ed. Olga Soffer and N. D. Praslov (New York: Plenum Press, 1993), 85–100; John F. Hoffecker, *Desolate Landscapes: Ice-Age Settlement in Eastern Europe* (New Brunswick, N.J.: Rutgers University Press, 2002), 206–232.

103. James R. Sackett, "The Neuvic Group: Upper Paleolithic Open-Air Sites in the Perigord," in *Upper Pleistocene Prehistory of Western Eurasia*, ed. Harold L. Dibble and Anna Montet-White (Philadelphia: University of Pennsylvania Museum, 1988), 61–84.

104. Hoffecker, "Representation and Recursion in the Archaeological Record," 380–381.

105. An exception is the hexagonal cell design of a honeybee comb. See, for example, Edward O. Wilson, *The Insect Societies* (Cambridge, Mass.: Belknap Press of Harvard University Press, 1971), 97, fig. 5–16.

106. Janusz K. Kozlowski, "The Gravettian in Central and Eastern Europe," in *Advances in World Archaeology*, ed. Fred Wendorf and Angela E. Close (Orlando, Fla.: Academic Press, 1986), 5:131–200; Svoboda, Ložek, and Vlček, *Hunters Between East and West*, 188–194.

5. Mindscapes of the Postglacial Epoch

1. The most impressive example of landscape modification in the Upper Paleolithic may be the introduction of one or more small mammals into New Ireland in the southwestern Pacific about 20,000 years ago. See Christopher Gosden, *Social Being and Time* (Oxford: Blackwell, 1994), 25.

2. André Leroi-Gourhan, *Gesture and Speech*, trans. Anna Bostock Berger (Cambridge, Mass.: MIT Press, 1993), 328–346.

3. Peter Gathercole, "Childe's Revolutions," in *Archaeology: The Key Concepts*, ed. Colin Renfrew and Paul Bahn (Abingdon: Routledge, 2005), 35–41.

4. Lewis Mumford, *The Myth of the Machine: Technics and Human Development* (New York: Harcourt, Brace & World, 1967), 190.

5. See, especially, Jan Assmann, *The Mind of Ancient Egypt: History and Meaning in the Time of the Pharaohs* (New York: Metropolitan Books, 2002), 46-49.

6. John D. Kasarda, "The Structural Implications of Social System Size: A Three-Level Analysis," *American Sociological Review* 39 (1974): 19-28, cited in the context of the rise of Sumerian civilization in Guillermo Algaze, *Ancient Mesopotamia at the Dawn of Civilization: The Evolution of an Urban Landscape* (Chicago: University of Chicago Press, 2008), 139.

7. Kwang-Chih Chang, *The Archaeology of Ancient China*, 4th ed. (New Haven, Conn.: Yale University Press, 1986); Michael D. Coe, *The Maya*, 6th ed. (New York: Thames and Hudson, 1999), 103-104; Charles Keith Maisels, *Early Civilizations of the Old World: The Formative Histories of Egypt, the Levant, Mesopotamia, India, and China* (London: Routledge, 1999).

8. The larger chiefdoms of Oceania are examples of large societies—perhaps tens of thousands of people—that did not become nation-states. See Patrick Vinton Kirch, *On the Road of the Winds: An Archaeological History of the Pacific Islands Before European Contact* (Berkeley: University of California Press, 2000), 311-313.

9. Barry J. Kemp, *Ancient Egypt: Anatomy of a Civilization* (London: Routledge, 1989), 19.

10. See, for example, Robert McC. Adams, *The Evolution of Urban Society: Early Mesopotamia and Prehispanic Mexico* (Chicago: Aldine, 1966). V. Gordon Childe enumerated ten criteria of civilizations (for example, monumental public buildings, writing) in "The Urban Revolution," *Town Planning Review* 21 (1950): 3-17. Childe also observed that "all measurement involves abstract thinking. In measuring lengths of stuffs you ignore their materials, colours, patterns, textures, and so on, to concentrate on length" (*Man Makes Himself* [London: Watts, 1936], 219).

11. According to Assmann, "The monumental tombs of Egypt are not graves in any contemporary sense. Their significance in Egyptian civilization is comparable to that which we attach to art and literature" (*Mind of Ancient Egypt*, 67).

12. Bert Hölldobler and E. O. Wilson, *The Super-Organism: The Beauty, Elegance, and Strangeness of Insect Societies* (New York: Norton, 2009).

13. The limited pace of invention during ancient times—compared with both the preceding and later epochs—has been discussed by many, including Childe, *Man Makes Himself*, 257-270; and Mumford, *Myth of the Machine*, 234-262. See also M. I. Finley, *The Ancient Economy*, 2nd ed., updated (Berkeley: University of California Press, 1999); Robert McC. Adams, *Paths of Fire: An Anthropologist's Inquiry into Western Technology* (Princeton, N.J.: Princeton University Press, 1996), 37-46; and Joel Mokyr, *The Lever of Riches: Technological Creativity and Economic Progress* (New York: Oxford University Press, 1990), 19-30.

14. Adams, *Paths of Fire*, 41.

15. In recent decades, historians of technology have modified the picture of stagnation in late antiquity—especially for the late Roman period—noting that innovations and applications of novel technologies (particularly water wheels) were more common than previously supposed. See, for example, Kevin Greene,

"Technological Innovation and Economic Progress in the Ancient World: M. I. Finley Reconsidered," *Economic History Review*, n.s., 53 (2000): 29–59; and Luke Lavan, "Explaining Technological Change: Innovation, Stagnation, Recession and Replacement," in *Technology in Transition*, A.D. *300–650*, ed. Luke Lavan, Enrico Zanini, and Alexander Sarantis (Leiden: Brill, 2007), xv–xl.

16. Claude Lévi-Strauss, *The Savage Mind* (Chicago: University of Chicago Press, 1966), 234. See also Roy A. Rappaport, *Ritual and Religion in the Making of Humanity* (Cambridge: Cambridge University Press, 1999), 17–22. Rappaport observed that "problems set by Alternatives arise, as much or more from the ordering of symbols through grammar . . . [which] makes the conception of alternatives virtually ineluctable" (17).

17. Peter Bellwood, *First Farmers: The Origins of Agricultural Societies* (Malden, Mass.: Blackwell, 2005), 44–49.

18. Dolores R. Piperno et al., "Processing of Wild Cereal Grains in the Upper Palaeolithic Revealed by Starch Grain Analysis," *Nature* 430 (2005): 670–673. The authors note ethnographic examples of baking ovens similar to the feature discovered at Ohalo II. Devices and procedures for the preparation of food have been termed "external digestion" and are another example of humans redesigning themselves as organisms with technology. See Martin Jones, "Moving North: Archaeobotanical Evidence for Plant Diet in Middle and Upper Paleolithic Europe," in *The Evolution of Hominin Diets: Integrating Approaches to the Study of Palaeolithic Subsistence*, ed. Jean-Jacques Hublin and Michael P. Richards (Dordrecht: Springer, 2009), 171–180.

19. Ofer Bar-Yosef, "The Natufian Culture in the Levant, Threshold to the Origins of Agriculture," *Evolutionary Anthropology* 6 (1998): 173.

20. Jean Perrot, "Le gisement natoufien de Mallaha (Eynan), Israel," *L'Anthropologie* 70 (1966): 437–484; Gordon C. Hillman, "Late Pleistocene Changes in Wild Plant Foods Available to Hunter-Gatherers of the Northern Fertile Crescent: Possible Preludes to Cereal Cultivation," in *The Origins and Spread of Agriculture and Pastoralism in Eurasia*, ed. David R. Harris (Washington, D.C.: Smithsonian Institution Press, 1996), 159–203. For a widely accessible account of the Natufian and the transition to sedentary life with an emphasis on the role of climate change, see Brian Fagan, *The Long Summer: How Climate Changed Civilization* (New York: Basic Books, 2004), 79–125.

21. A. M. T. Moore, G. C. Hillman, and A. J. Legge, *Village on the Euphrates: From Foraging to Farming at Abu Hureyra* (London: Oxford University Press, 2000); Bar-Yosef, "Natufian Culture in the Levant," 172–174; Bellwood, *First Farmers*, 49–54.

22. Gordon C. Hillman and M. Stuart Davies, "Measured Domestication Rates in Wild Wheat and Barley Under Primitive Cultivation and Their Archaeological Implications," *Journal of World Prehistory* 4 (1990): 157–222.

23. Bellwood, *First Farmers*, 59–66.

24. Robert McC. Adams and Hans J. Nissen, *The Uruk Countryside: The Natural Setting of Urban Societies* (Chicago: University of Chicago Press, 1972); Guillermo Algaze, *The Uruk World System: The Dynamics of Expansion of Early Mesopotamian Civilization*, 2nd ed. (Chicago: University of Chicago Press, 2005).

25. Robert McC. Adams, *Heartland of Cities: Surveys of Ancient Settlement and Land Use on the Central Floodplain of the Euphrates* (Chicago: University of Chicago Press, 1981); Maisels, *Early Civilizations of the Old World*, 175.

26. See the classic Samuel Noah Kramer, *The Sumerians: Their History, Culture, and Character* (Chicago: University of Chicago Press, 1963). Sumer did not become a unified nation-state for another 700 years, when the southern cities were unified under Lugal-zagesi in 2340 B.C.E. See Georges Roux, *Ancient Iraq*, 3rd ed. (London: Penguin, 1992).

27. Frank Hole, "Environmental Instabilities and Urban Origins," in *Chiefdoms and Early States in the Near East: The Organizational Dynamics of Complexity*, ed. Gil Stein and Mitchell S. Rothman (Madison, Wis.: Prehistory Press, 1994), 121–143. See also Fagan, *Long Summer*, 134–139. Algaze wrote, "Early Near Eastern villagers domesticated plants and animals. Uruk urban institutions, in turn, domesticated humans" (*Ancient Mesopotamia at the Dawn of Civilization*, 129).

28. Kramer, *Sumerians*, 77–88; Maisels, *Early Civilizations of the Old World*, 166–169.

29. Algaze, *Ancient Mesopotamia at the Dawn of Civilization*, 135–137.

30. Ibid., 66–68; Kramer, *Sumerians*, 104. A model sailboat was recovered from Eridu.

31. V. Gordon Childe, "Wheeled Vehicles," in *A History of Technology*, vol. 1, *From Early Times to the Fall of Ancient Empires*, ed. Charles Singer, Eric J. Holmyard, and A. R. Hall (Oxford: Clarendon Press, 1954), 716–729; Kramer, *Sumerians*, 104–105; L. Sprague de Camp, *The Ancient Engineers* (New York: Ballantine Books, 1963), 57–58; Glyn Daniel, *The First Civilizations: The Archaeology of Their Origins* (New York: Crowell, 1968), 72–73; Stuart Piggott, *The Earliest Wheeled Transport: From the Atlantic Coast to the Caspian Sea* (London: Thames and Hudson, 1983).

32. The true circle itself is not found in the natural world, and is a concept generated by the modern human mind, according to Ernest Zebrowski, *A History of the Circle: Mathematical Reasoning and the Physical Universe* (New Brunswick, N.J.: Rutgers University Press, 1999).

33. Algaze, *Ancient Mesopotamia at the Dawn of Civilization*, 131.

34. Kramer, *Sumerians*, 104; Algaze, *Ancient Mesopotamia at the Dawn of Civilization*, 82–83.

35. My own view is that the engineering of the public buildings is far outweighed in originality and significance by the innovations in information technology and applications of animal and wind power in agriculture and transportation.

36. Kramer, *Sumerians*, 135–144.

37. I. E. S. Edwards, *The Pyramids of Egypt* (Harmondsworth: Penguin, 1961), 116–138. See also Mumford, *Myth of the Machine*, 194–198.

38. Kramer, *Sumerians*, 91; Maisels, *Early Civilizations of the Old World*, 74–75.

39. Kemp, *Ancient Egypt*, 31–35. For a comparison of Sumer and Egypt, see Maisels, *Early Civilizations of the Old World*.

40. Chang, *Archaeology of Ancient China*, 309–331. Large structures were constructed during the Shang dynasty, however, including the palace at Erh-li-t'ou and the subterranean royal tombs at An-yang. Regarding the foundation of the state in China, Maisels wrote: "The original pre-state solidarity was the highly potent

solidarity of the clan. . . . Social divisions were enhanced with the state, but as its advent was an incremental process and not a revolutionary one" (*Early Civilizations of the Old World*, 339–340).

41. Coe, *Maya*, 41–55.
42. Ibid., 103–104.
43. Marcia Ascher and Robert Ascher, *Code of the Quipu: A Study in Media, Mathematics, and Culture* (Ann Arbor: University of Michigan Press, 1980).
44. R. G. Collingwood, *The Idea of History* (Oxford: Clarendon Press, 1946), 232.
45. Herbert Butterfield, *The Origins of History* (New York: Basic Books, 1981). See also Assmann, *Mind of Ancient Egypt*, 13–17.
46. Claude Lévi-Strauss, *Structural Anthropology*, trans. Claire Jacobson and Brooke Grundfest Schoepf (New York: Basic Books, 1963), 209.
47. Claude Lévi-Strauss, *The Raw and the Cooked: Introduction to a Science of Mythology*, trans. John Weightman and Doreen Weightman (New York: Harper & Row, 1969), 1:16. Edmund Leach translated the original French text as "machines for the suppression of time" (*Claude Lévi-Strauss* [New York: Viking Press, 1970], 125).
48. Collingwood, *Idea of History*, 14–17; Butterfield, *Origins of History*, 22–79.
49. Assmann, *Mind of Ancient Egypt*, 20–21. Many aspects of Egyptian technology are described in A. Lucas and J. R. Harris, *Ancient Egyptian Materials and Industries*, 4th ed. (London: Arnold, 1962).
50. The *sed* festival was performed when the pharaoh reached the thirtieth year of his reign. According to Assmann, "Djoser's casting of the ritual structures in stone rather than the transient materials previously employed served the purpose of enabling the king to continue the sed festival into all eternity" (*Mind of Ancient Egypt*, 56).
51. Assmann wrote, "Where ritual was concerned . . . [e]verything hinged on precise reiteration. Maximum care was taken to prevent deviation and improvisation" (ibid., 71). The irony of these observations is that historians traditionally have viewed the invention of writing as the beginning of history (describing preliterate societies as "people without history"), while it would seem that the Egyptians used writing to help suppress history.
52. Assmann noted that "the division into nomes represented a thoroughgoing reorganization of the territory, undertaken probably as late as the reign of Djoser (2687–2667 B.C.). There is thus no continuity between the rival chiefdoms of the Naqada Period and the nomes of the Old Kingdom. Indeed, the structures that had evolved before the advent of the state were ruthlessly suppressed" (ibid., 47).
53. According to Assmann, "The police state character of the Middle Kingdom is the inevitable institutional expression of a state that styles itself primarily as a bulwark against chaos, as a bastion of a civilization built upon law, order, and justice. Such a state will inevitably develop organs of control, surveillance, and punishment that curtail individual freedom of movement" (ibid., 139).
54. Maisels, *Early Civilizations of the Old World*, 185; Algaze, *Ancient Mesopotamia at the Dawn of Civilization*, 128.
55. de Camp, *Ancient Engineers*, 57–82; John McLeish, *The Story of Numbers: How Mathematics Shaped Civilization* (New York: Fawcett Columbine, 1991), 32–37; David C.

Lindberg, *The Beginnings of Western Science: The European Scientific Tradition in Philosophical, Religious, and Institutional Context, 600* B.C. *to* A.D. *1450* (Chicago: University of Chicago Press, 1992), 13–20. See also Roux, *Ancient Iraq*, 357–366.

56. J. R. McNeill and William H. McNeill, *The Human Web: A Bird's-Eye View of World History* (New York: Norton, 2003), 55–82; Daniel R. Headrick, *Technology: A World History* (Oxford: Oxford University Press, 2009), 35–50. No fewer than three revolutions in military technology took place before 500 B.C.E., and all of them seem to have emerged from the uncivilized fringe. The first was the "chariot revolution" at about 1700 B.C.E., which overran both Mesopotamia and Egypt and eventually hit China.

57. For a recent discussion of the Hellenistic "military revolution," see S. Cuomo, *Technology and Culture in Greek and Roman Antiquity* (Cambridge: Cambridge University Press, 2007), 41–76. On mechanical technology, see Andrew I. Wilson, "Machines in Greek and Roman Technology," in *The Oxford Handbook of Engineering and Technology in the Classical World*, ed. John Peter Oleson (Oxford: Oxford University Press, 2008), 336–366.

58. de Camp, *Ancient Engineers*, 83–171; Geoffrey Lloyd, *Greek Science After Aristotle* (London: Chatto and Windus, 1973); J. G. Landels, *Engineering in the Ancient World* (Berkeley: University of California Press, 1978); Alan Hirshfeld, *Eureka Man: The Life and Legacy of Archimedes* (New York: Walker, 2009). One of the most spectacular examples of Hellenistic technology is a device known as the Antikythera Mechanism, which was recovered from a sunken ship in the Aegean Sea in 1900 and is thought to date to the mid-first century B.C.E. at the latest. For a discussion of recent research on this artifact, described as a mechanical "planetarium," and a depiction of a working model, see Robert Hannah, "Timekeeping," in *Oxford Handbook of Engineering and Technology in the Classical World*, ed. Oleson, 744–746. The attribution of the crank to Hellenistic engineers is problematic, however, according to Lynn White, *Medieval Technology and Social Change* (London: Oxford University Press, 1962), 103–115. For the link between Archimedes and Galileo, see Donald Cardwell, *The Norton History of Technology* (New York: Norton, 1995), 83–84.

59. de Camp, *Ancient Engineers*, 130–138; Cardwell, *Norton History of Technology*, 20–24. Fernand Braudel discussed Alexandrian Egypt as an abortive industrial revolution in *The Perspective of the World*, vol. 3 of *Civilization and Capitalism, 15th–18th Century*, trans. Siân Reynolds (Berkeley: University of California Press, 1981), 543–544.

60. See, for example, Charles Freeman, *The Greek Achievement: The Foundation of the Western World* (New York: Viking Press, 1999), 372–388.

61. Chang concluded that "no significantly new technological invention has been archaeologically documented from the Neolithic into the Bronze Age. . . . The emergence of Bronze Age civilizations in China was not accompanied . . . by a significant use of metal farming implements, irrigation networks, any use of draft animals, or the use of the plough" (*Archaeology of Ancient China*, 364).

62. Kwang-Chih Chang, *The Archaeology of Ancient China*, 3rd ed. (New Haven, Conn.: Yale University Press, 1977), 350–357; Robert Temple, *The Genius of China: 3,000 Years of Science, Discovery, and Invention* (New York: Simon and Schuster, 1986), 15–23; Headrick, *Technology*, 52–53.

63. Mark Elvin, *The Pattern of the Chinese Past* (Stanford, Calif.: Stanford University Press, 1973), 23–34.

64. Joseph Needham, *The Development of Iron and Steel Technology in China* (Cambridge: Heffer, 1964); William H. McNeill, *The Pursuit of Power: Technology, Armed Force, and Society Since A.D. 1000* (Chicago: University of Chicago Press, 1982), 24–62; Arnold Pacey, *Technology in World Civilization: A Thousand-Year History* (Cambridge, Mass.: MIT Press, 1990), 1–6.

65. Joseph Needham, *Science in Traditional China* (Cambridge, Mass.: Harvard University Press, 1981). See also Temple, *Genius of China*, 103–110, which depicts a model of Su Sung's water-powered mechanical clock of 1092 (fig. 79). For a description of Chinese nautical technology, including watertight compartments and a magnetic compass, see Joseph Needham, *The Shorter Science and Civilization in China*, vol. 3 (Cambridge: Cambridge University Press, 1986).

66. Elvin, *Pattern of the Chinese Past*, 91–110. The Chinese Communist Party seized control of the nation in 1949—after years of foreign invasion and civil war—in much the same manner as had earlier dynasties, and it explicitly emulated the Qin dynasty in expanding the power and reach of the central government in concert with a state ideology. In recent decades, the Chinese leaders have loosened control of the economy to permit more rapid growth and have followed (less explicitly) the course of other dynasties like the Zhou and Han that exerted less rigid control over the provinces and cities.

67. McNeill, *Pursuit of Power*, 44–50; Mokyr, *Lever of Riches*, 231–238; Headrick, *Technology*, 72–73.

68. Jean Gimpel, *The Medieval Machine: The Industrial Revolution of the Middle Ages* (New York: Holt, 1976), 29–58; Fernand Braudel, *The Structures of Everyday Life: The Limits of the Possible*, vol. 1 of *Civilization and Capitalism, 15th–18th Century*, trans. Siân Reynolds (Berkeley: University of California Press, 1981), 104–145.

69. McNeill, *Pursuit of Power*, 63–116; Mokyr, *Lever of Riches*, 231–238; David S. Landes, *The Wealth and Poverty of Nations: Why Some Are So Rich and Some So Poor* (New York: Norton, 1999), 29–59.

70. Ernst Benz, *Evolution and Christian Hope: Man's Concept of the Future from the Early Fathers to Teilhard de Chardin*, trans. Heinz G. Frank (Garden City, N.Y.: Doubleday, 1966); Lynn White, *Medieval Religion and Technology: Collected Essays* (Berkeley: University of California Press, 1978). See also Mokyr, *Lever of Riches*, 201–208.

71. Cardwell describes the attitude toward manual labor and technical matters in *Norton History of Technology*, 37. See also Mumford, *Myth of the Machine*, 263–267. For a discussion of the Judeo-Christian construction of time, see J. B. Bury, *The Idea of Progress: An Inquiry into Its Origin and Growth* (New York: Macmillan, 1932), 20–29. Butterfield considers the importance of Saint Augustine's writings on time and history in *Origins of History*, 180–184.

72. White, *Medieval Technology and Social Change*, 39–78; Gimpel, *Medieval Machine*, 32–43; Headrick, *Technology*, 55–58.

73. Mumford, *Myth of the Machine*, 234–262. Although Mumford expressed skepticism about the decline of slavery as a significant factor in accelerated technological innovation, Gimpel noted a general correlation between the downward trend in slavery and the increase in water power in *Medieval Machine*, 10.

74. Landels, *Engineering in the Ancient World*, 16–26. Headrick described the Roman hydropower complex at Barbegal in France (constructed in 310 C.E.), which comprised sixteen overshot wheels and was capable of grinding 3 tons of grain an hour, in *Technology*, 49–50.

75. White, *Medieval Technology and Social Change*, 79–85; Gimpel, *Medieval Machine*, 1–24.

76. Pacey, *Technology in World Civilization*, 10–12.

77. White, *Medieval Technology and Social Change*, 84–89; Gimpel, *Medieval Machine*, 24–27. See also Lewis Mumford, *Technics and Civilization* (New York: Harcourt, Brace, 1934), 112–118. Windmills were so common by the last decade of the twelfth century that they were taxed by Pope Celestine III.

78. Mumford, *Technics and Civilization*, 132–134; David S. Landes, *Revolution in Time: Clocks and the Making of the Modern World* (Cambridge, Mass.: Harvard University Press, 1983), 53–66.

79. Arnold Pacey, *The Maze of Ingenuity: Ideas and Idealism in the Development of Technology*, 2nd ed. (Cambridge, Mass.: MIT Press, 1992), 35–45.

80. Landes, *Revolution in Time*, 57.

81. Mumford, *Technics and Civilization*, 126; Landes, *Wealth and Poverty of Nations*, 46–47.

82. Gimpel, *Medieval Machine*, 199–236. For an accessible account of this period, see Barbara W. Tuchman, *A Distant Mirror: The Calamitous 14th Century* (New York: Knopf, 1978).

83. Cardwell, *Norton History of Technology*, 49–56.

84. Braudel, *Structures of Everyday Life*, 385–397; McNeill, *Pursuit of Power*, 79–143.

85. Braudel, *Structures of Everyday Life*, 402–415.

86. Mumford, *Technics and Civilization*, 46.

87. See, for example, Cardwell, *Norton History of Technology*, 75–101. Contributions came from other parts of Europe as well, most notably from Nicolaus Copernicus (1473–1543) in Poland.

88. Gimpel, *Medieval Machine*, 149.

89. Bury, *Idea of Progress*, 50–63.

90. For a brief but accessible overview, see Daniel J. Boorstin, *The Discoverers: A History of Man's Search to Know His World and Himself* (New York: Random House, 1983), 626–652.

91. William J. Sollas, *Ancient Hunters and Their Modern Representatives* (London: Macmillan, 1911). For a readable account of the European encounter with the native Australians, see Alan Moorehead, *The Fatal Impact: An Account of the Invasion of the South Pacific, 1767–1840* (New York: Harper & Row, 1966).

92. Marvin Harris, *The Rise of Anthropological Theory: A History of Theories of Culture* (New York: Crowell, 1968), 149–179; Robert L. Kelly, *The Foraging Spectrum: Diversity in Hunter-Gatherer Lifeways* (Washington, D.C.: Smithsonian Institution Press, 1995), 6–10.

93. See, for example, Harris, *Rise of Anthropological Theory*, 80–107; and Bruce G. Trigger, *A History of Archaeological Thought* (Cambridge: Cambridge University Press, 1989), 111–114.

94. Kelly, *Foraging Spectrum*.

95. Peter Hiscock concluded that "economic, social, and ideological change was not restricted to the historical period or even recent millennia, but occurred throughout Australian pre-history. . . . [T]here were ongoing modifications to foraging practices, technology, settlement and territoriality, and to social practices and the nature of cosmology and belief" (*Archaeology of Ancient Australia* [London: Routledge, 2008], 284–285). See also Harry Lourandos, *Continent of Hunter-Gatherers: New Perspectives in Australian Prehistory* (Cambridge: Cambridge University Press, 1997).

96. For example, Wendell H. Oswalt observed that the Aranda (Arunta), often considered the most technologically simple of the native Australian groups, manufactured a throwing board spear composed of nine "techno-units," or components, as well as untended facility in the form of an emu pit trap of four components (*An Anthropological Analysis of Food-Getting Technology* [New York: Wiley, 1976], 236–237).

97. Robert McGhee, *Ancient People of the Arctic* (Vancouver: University of British Columbia Press, 1996).

6. The Vision Animal

1. Edward O. Wilson, *The Insect Societies* (Cambridge, Mass.: Belknap Press of Harvard University Press, 1971); Bert Hölldobler and E. O. Wilson, *The Super-Organism: The Beauty, Elegance, and Strangeness of Insect Societies* (New York: Norton, 2009), 5–10.

2. Hölldobler and Wilson, *Super-Organism*, 408–467.

3. The eusocial rodents include the naked mole-rat (*Heterocephalus glaber*) and the Damaraland mole-rat (*Cryptomys damarensis*). See J. U. M. Jarvis, "Eusociality in a Mammal: Cooperative Breeding in Naked Mole-Rat Colonies," *Science* 212 (1981): 571–573; and M. Andersson, "The Evolution of Eusociality," *Annual Reviews of Ecological Systems* 15 (1984): 165–189.

4. See, for example, George Gaylord Simpson, *The Major Features of Evolution* (New York: Simon and Schuster, 1953), 177–179; and Niles Eldredge, *Macro-Evolutionary Dynamics: Species, Niches, and Adaptive Peaks* (New York: McGraw-Hill, 1989), 51–53.

5. See, for example, Ludwig von Bertalanffy, *General System Theory: Foundations, Development, Applications* (New York: Braziller, 1968). The study of complex biological systems is termed "bio-cybernetics."

6. Hölldobler and Wilson, *Super-Organism*, 4–13. See also Douglas R. Hofstadter, *Gödel, Escher, Bach: An Eternal Golden Braid* (New York: Basic Books, 1979), 358–361.

7. R. G. Collingwood, *The Idea of History* (Oxford: Clarendon Press, 1946), 88–93; Marvin Harris, *The Rise of Anthropological Theory: A History of Theories of Culture* (New York: Crowell, 1968), 80–141.

8. Collingwood, *Idea of History*, 209.

9. Bruce G. Trigger, "Childe's Relevance to the 1990s," in *The Archaeology of V. Gordon Childe*, ed. David R. Harris (Chicago: University of Chicago Press, 1994), 9–27.

220   6. THE VISION ANIMAL

10. Collingwood, *Idea of History*, 190–204.

11. Immanuel Kant, *On History*, ed. Lewis White Beck, trans. Lewis White Beck, Robert E. Anchor, and Emil L. Fackenheim (New York: Macmillan, 1963), 12.

12. Collingwood, *Idea of History*, 97. Concerning the collective and cumulative character of knowledge, Kant wrote that "a single man would have to live excessively long in order to learn to make full use of all his natural capacities. Since Nature has set only a short period for his life, she needs a perhaps unreckonable series of generations, each of which passes its own enlightenment to its successor in order finally to bring the seeds of enlightenment to that degree of development in our race which is completely suitable to Nature's purpose" ("Idea for a Universal History from a Cosmopolitan Point of View," trans. Beck, in *On History*, 13).

13. Kant's view of history was presented in his short essay "Idea for a Universal History from a Cosmopolitan Point of View" (1784), published in *Gothaische Gelehrte Zeitung* and reprinted in Kant, *On History*, 11–26. Kant was by no means certain that progress would continue; he suggested that future progress might be obliterated by "barbarous devastation," which seemed to be an eerie forecast of the massive military–industrial conflicts of the twentieth century.

14. Georg Wilhelm Friedrich Hegel, *Philosophy of History*, trans. J. Sibree (New York: Barnes & Noble, 2004), 59. Hegel's distinction between history and Nature was based on an erroneous view of the latter as static.

15. Ibid., 30

16. Collingwood, *Idea of History*, 113–122.

17. William H. McNeill, *The Pursuit of Power: Technology, Armed Force, and Society Since A.D. 1000* (Chicago: University of Chicago Press, 1982), 63–143.

18. Mary Shelley described the genesis of her novel in the well-known author's introduction to *Frankenstein* (New York: Random House, 1993), xiii–xxi.

19. Ibid., xix.

20. Robert Hannah, "Timekeeping," in *The Oxford Handbook of Engineering and Technology in the Classical World*, ed. John Peter Oleson (Oxford: Oxford University Press, 2008), 744–746; Donald Cardwell, *The Norton History of Technology* (New York: Norton, 1995), 420–422.

21. Margaret A. Boden, *Computer Models of Mind: Computational Approaches in Theoretical Psychology* (Cambridge: Cambridge University Press, 1988), and *The Creative Mind: Myths and Mechanisms*, 2nd ed. (London: Routledge, 2004).

22. Alan Turing, "Computing Machinery and Intelligence," *Mind* 59 (1950): 433–460. Turing's criterion for AI has long been referred to as the "Turing test."

23. Ray Kurzweil, *The Singularity Is Near: When Humans Transcend Biology* (New York: Viking Press, 2005), 136. See also Ray Kurzweil, *The Age of Spiritual Machines: When Computers Exceed Human Intelligence* (New York: Viking Press, 1999).

24. K. Eric Drexler, *Engines of Creation: The Coming Era of Nanotechnology* (New York: Anchor Books, 1986).

25. Kurzweil, *Singularity Is Near*, 299–367.

26. George B. Dyson, *Darwin Among the Machines: The Evolution of Global Intelligence* (Reading, Mass.: Addison-Wesley, 1997), 211–228.

BIBLIOGRAPHY

Abramova, Zoya A. "Two Examples of Terminal Paleolithic Adaptations." In *From Kostenki to Clovis: Upper Paleolithic–Paleo-Indian Adaptations*, edited by Olga Soffer and N. D. Praslov, 85–100. New York: Plenum Press, 1993.

Adams, Robert McC. *The Evolution of Urban Society: Early Mesopotamia and Prehispanic Mexico*. Chicago: Aldine, 1966.

———. *Heartland of Cities: Surveys of Ancient Settlement and Land Use on the Central Floodplain of the Euphrates*. Chicago: University of Chicago Press, 1981.

———. *Paths of Fire: An Anthropologist's Inquiry into Western Technology*. Princeton, N.J.: Princeton University Press, 1996.

Adams, Robert McC., and Hans J. Nissen. *The Uruk Countryside: The Natural Setting of Urban Societies*. Chicago: University of Chicago Press, 1972.

Aiello, Leslie, and Christopher Dean. *An Introduction to Human Evolutionary Anatomy*. London: Academic Press, 1990.

Algaze, Guillermo. *Ancient Mesopotamia at the Dawn of Civilization: The Evolution of an Urban Landscape*. Chicago: University of Chicago Press, 2008.

———. *The Uruk World System: The Dynamics of Expansion of Early Mesopotamian Civilization*. 2nd ed. Chicago: University of Chicago Press, 2005.

Allman, John Morgan. *Evolving Brains*. New York: Scientific American Library, 1999.

Allsworth-Jones, Philip. "The Szeletian and the Stratigraphic Succession in Central Europe and Adjacent Areas: Main Trends, Recent Results, and Problems for Resolution." In *The Emergence of Modern Humans: An Archaeological Perspective*, edited by Paul Mellars, 160–242. Edinburgh: Edinburgh University Press, 1990.

Ambrose, Stanley H. "Late Pleistocene Human Population Bottlenecks, Volcanic Winter, and Differentiation of Modern Humans." *Journal of Human Evolution* 34 (1998): 623–651.

———. "Paleolithic Technology and Human Evolution." *Science* 291 (2001): 1748–1753.

Anderson, Michael L. "Embodied Cognition: A Field Guide." *Artificial Intelligence* 149 (2003): 91–130.

Anderson-Gerfaud, Patricia. "Aspects of Behavior in the Middle Palaeolithic: Functional Analysis of Stone Tools from Southwest France." In *The Emergence of Modern Humans: An Archaeological Perspective*, edited by Paul Mellars, 389–418. Edinburgh: Edinburgh University Press, 1990.

Andersson, Malte. "The Evolution of Eusociality." *Annual Reviews of Ecological Systems* 15 (1984): 165–189.

Ascher, Marcia, and Robert Ascher. *Code of the Quipu: A Study in Media, Mathematics, and Culture*. Ann Arbor: University of Michigan Press, 1980.

Asfaw, Berhane, Yonas Beyene, Gen Suwa, Robert C. Walter, Tim D. White, Giday WoldeGabriel, and Tesfaye Yemane. "The Earliest Acheulean from Konso-Gardula." *Nature* 360 (1992): 732–735.

Assmann, Jan. *The Mind of Ancient Egypt: History and Meaning in the Time of the Pharaohs*. New York: Metropolitan Books, 2002.

Bailey, Shara E., Timothy D. Weaver, and Jean-Jacques Hublin. "Who Made the Aurignacian and Other Early Upper Paleolithic Industries?" *Journal of Human Evolution* 57 (2009): 11–26.

Barkow, Jerome H., Leda Cosmides, and John Tooby, eds. *The Adapted Mind: Evolutionary Psychology and the Generation of Culture*. New York: Oxford University Press, 1992.

Bar-Mathews, Miryam, and Avner Ayalon. "Climatic Conditions in the Eastern Mediterranean During the Last Glacial (60–10 ky) and Their Relations to the Upper Palaeolithic in the Levant as Inferred from Oxygen and Carbon Isotope Systematics of Cave Deposits." In *More Than Meets the Eye: Studies on Upper Palaeolithic Diversity in the Near East*, edited by A. Nigel Goring-Morris and Anna Belfer-Cohen, 13–18. Oxford: Oxbow Books, 2003.

Bar-Yosef, Ofer. "The Middle and Early Upper Paleolithic in Southwest Asia and Neighboring Regions." In *The Geography of Neandertals and Modern Humans in Europe and the Greater Mediterranean*, edited by Ofer Bar-Yosef and David Pilbeam, 107–156. Cambridge, Mass.: Peabody Museum of Archaeology and Ethnography, 2000.

———. "The Natufian Culture in the Levant, Threshold to the Origins of Agriculture." *Evolutionary Anthropology* 6 (1998): 159–177.

Bar-Yosef, Ofer, and Anna Belfer-Cohen. "From Africa to Eurasia—Early Dispersals." *Quaternary International* 75 (2001): 19–28.

Becker, Ernest. *The Denial of Death*. New York: Free Press, 1973.

Bell, Charles. *The Hand: Its Mechanism and Vital Endowments as Evincing Design*. New York: Harper, 1840.

Bellwood, Peter. *First Farmers: The Origins of Agricultural Societies*. Malden, Mass.: Blackwell, 2005.

Benecke, Norbert. "Studies on Early Dog Remains from Northern Europe." *Journal of Archaeological Science* 14 (1987): 31–49.

Benz, Ernst. *Evolution and Christian Hope: Man's Concept of the Future from the Early Fathers to Teilhard de Chardin*. Translated by Heinz G. Frank. Garden City, N.Y.: Doubleday, 1966.

Bertalanffy, Ludwig von. *General System Theory: Foundations, Development, Applications*. New York: Braziller, 1968.

Beyries, Sylvia. "Functional Variability of Lithic Sets in the Middle Paleolithic." In *Upper Pleistocene Prehistory of Western Eurasia*, edited by Harold L. Dibble and Anta Montet-White, 213–224. Philadelphia: University of Pennsylvania Museum, 1988.

Bickerton, Derek. *Language and Human Behavior*. Seattle: University of Washington Press, 1995.

———. *Language and Species*. Chicago: University of Chicago Press, 1990.

Binford, Lewis R., and Sally R. Binford. "A Preliminary Analysis of Functional Variability in the Mousterian of Levallois Facies." *American Anthropologist* 68 (1966): 238–295.

Boaretto, Elisabetta, Xiaohong Wu, Jiarong Yuan, Ofer Bar-Yosef, Vikki Chu, Yan Pan, Kexin Lu, et al. "Radiocarbon Dating of Charcoal and Bone Collagen Associated with Early Pottery at Yuchanyan Cave, Hunan Province, China." *Proceedings of the National Academy of Sciences* 106 (2009): 9595–9600.

Boaz, Noel T., and Russell L. Ciochon. *Dragon Bone Hill: An Ice-Age Saga of Homo erectus*. Oxford: Oxford University Press, 2004.

Bochenski, Zbigniew M., Teresa Tomek, Jarosław Wilczyński, Jiří Svoboda, Krzysztof Wertz, and Piotr Wojtal. "Fowling During the Gravettian: The Avifauna of Pavlov I, Czech Republic." *Journal of Archaeological Science* 36 (2009): 2655–2665.

Bocherens, Hervé, Dorothée G. Drucker, Daniel Billiou, Marylène Patou-Mathis, and Bernard Vandermeersch. "Isotopic Evidence for Diet and Subsistence Pattern of the Saint Cesaire I Neanderthal: Review and Use of a Multi-source Mixing Model." *Journal of Human Evolution* 49 (2005): 71–87.

Boden, Margaret A. *Computer Models of Mind: Computational Approaches in Theoretical Psychology*. Cambridge: Cambridge University Press, 1988.

———. *The Creative Mind: Myths and Mechanisms*. 2nd ed. London: Routledge, 2004.

Boëda, Eric. "Levallois: A Volumetric Construction, Methods, a Technique." In *The Definition and Interpretation of Levallois Technology*, edited by Harold L Dibble and Ofer Bar-Yosef, 41–68. Madison, Wis.: Prehistory Press, 1995.

Boëda, Eric, Jacques Connan, Daniel Dessort, Sultan Muhesen, Norbert Mercier, Hélène Valladas, and Nadine Tisnérat. "Bitumen as a Hafting Material on Middle Palaeolithic Artefacts." *Nature* 380 (1996): 336–338.

Bonin, Gerhardt von. *The Evolution of the Human Brain*. Chicago: University of Chicago Press, 1963.

Boorstin, Daniel J. *The Discoverers: A History of Man's Search to Know His World and Himself*. New York: Random House, 1983.

Bordes, François. "Mousterian Cultures in France." *Science* 134 (1961): 803–810.

———. *Typologie du Paléolithique ancient et moyen*. Bordeaux: Imprimeries Delmas, 1961.

Braudel, Fernand. *The Perspective of the World*. Vol. 3 of *Civilization and Capitalism, 15th–18th Century*. Translated by Siân Reynolds. Berkeley: University of California Press, 1981.

——. *The Structures of Everyday Life: The Limits of the Possible*. Vol. 1 of *Civilization and Capitalism, 15th–18th Century*. Translated by Siân Reynolds. Berkeley: University of California Press, 1981.

Brunet, Michel, Franck Guy, David Pilbeam, Hassane Taisso Mackaye, Andossa Likius, Djimdoumalbaye Ahounta, Alain Beauvilain, et al. "A New Hominid from the Upper Miocene of Chad, Central Africa." *Nature* 418 (2002): 145–801.

Bullock, Theodore Holmes, and G. Adrian Horridge. *Structure and Function in the Nervous Systems of Invertebrates*. San Francisco: Freeman, 1965.

Bunge, Mario. *Emergence and Convergence: Qualitative Novelty and the Unity of Knowledge*. Toronto: University of Toronto Press, 2003.

——. *The Mind–Body Problem: A Psychobiological Approach*. Oxford: Pergamon Press, 1980.

Bunn, Henry T., and Ellen M. Kroll. "Systematic Butchery by Plio/Pleistocene Hominids at Olduvai Gorge, Tanzania." *Current Anthropology* 27 (1986): 431–452.

Bury, J. B. *The Idea of Progress: An Inquiry into Its Origin and Growth*. New York: Macmillan, 1932.

Bush, Michael E., C. Owen Lovejoy, Donald C. Johanson, and Yves Coppens. "Hominid Carpal, Metacarpal, and Phalangeal Bones Recovered from the Hadar Formation: 1974–1977 Collections." *American Journal of Physical Anthropology* 57 (1982): 651–677.

Butler, Samuel. *The Shrewsbury Edition of the Works of Samuel Butler*. Vol. 1. Edited by Henry Festing Jones. London: Cape, 1923.

Butterfield, Herbert. *The Origins of History*. New York: Basic Books, 1981.

Byrne, Richard, and Andrew Whiten, eds. *Machiavellian Intelligence: Social Expertise and the Evolution of Intellect in Monkeys, Apes, and Humans*. Oxford: Clarendon Press, 1988.

Call, Josep, and Michael Tomasello. *The Gestural Communication of Apes and Monkeys*. Mahwah, N.J.: Erlbaum, 2007.

Callow, P. "The Lower and Middle Palaeolithic of Britain and Adjacent Areas of Europe." Ph.D. diss., Cambridge University, 1976.

Calvin, William H. "The Unitary Hypothesis: A Common Neural Circuitry for Novel Manipulations, Language, Plan-Ahead, and Throwing?" In *Tools, Language, and Cognition in Human Evolution*, edited by Kathleen R. Gibson and Tim Ingold, 230–250. Cambridge: Cambridge University Press, 1993.

Cardwell, Donald. *The Norton History of Technology*. New York: Norton, 1995.

Cartmill, Matt. "Rethinking Primate Origins." *Science* 184 (1974): 436–443.

Cattelain, Pierre. "Un crochet de propulseur solutréen de la grotte de Combe-Saunière 1 (Dordogne)." *Bulletin de la Société préhistorique française* 86 (1989): 213–216.

Chang, Kwang-Chih. *The Archaeology of Ancient China*. 3rd ed. New Haven, Conn.: Yale University Press, 1977.

——. *The Archaeology of Ancient China*. 4th ed. New Haven, Conn.: Yale University Press, 1986.

Cheney, Dorothy L., and Robert S. Seyfarth. *How Monkeys See the World: Inside the Mind of Another Species*. Chicago: University of Chicago Press, 1990.

Childe, V. Gordon. *The Dawn of European Civilization*. London: Kegan Paul, 1925.

——. *History*. London: Cobbett Press, 1947.

——. *Man Makes Himself*. London: Watts, 1936.

——. *New Light on the Most Ancient East: The Oriental Prelude to European Prehistory*. London: Kegan Paul, 1934.

——. *Piecing Together the Past: The Interpretation of Archaeological Data*. New York: Praeger, 1956.

——. *The Prehistory of European Society*. Harmondsworth: Penguin, 1958.

——. "Races, Peoples, and Cultures in Prehistoric Europe." *History* 18 (1933): 193–203.

——. "Retrospect." *Antiquity* 32 (1958): 69–74.

——. *Scotland Before the Scots*. London: Methuen, 1946.

——. *Social Evolution*. London: Watts, 1951.

——. *Society and Knowledge: The Growth of Human Traditions*. New York: Harper, 1956.

——. "The Urban Revolution." *Town Planning Review* 21 (1950): 3–17.

——. *What Happened in History*. Harmondsworth: Penguin, 1942.

——. "Wheeled Vehicles." In *A History of Technology*. Vol. 1, *From Early Times to the Fall of Ancient Empires*, edited by Charles Singer, Eric J. Holmyard, and A. R. Hall, 716–729. Oxford: Clarendon Press, 1954.

Chomsky, Noam. *Language and Mind*. 3rd ed. Cambridge: Cambridge University Press, 2006.

——. *Language and the Problems of Knowledge: The Managua Lectures*. Cambridge, Mass.: MIT Press, 1988.

——. *The Minimalist Program*. Cambridge, Mass.: MIT Press, 1995.

——. *On Nature and Language*. Cambridge: Cambridge University Press, 2002.

Christiansen, Morten H., and Nick Chater. "Language as Shaped by the Brain." *Behavioral and Brain Sciences* 31 (2008): 489–558.

Clark, Grahame. *Prehistoric Europe: The Economic Basis*. London: Methuen, 1952.

——. *World Prehistory in New Perspective*. Cambridge: Cambridge University Press, 1977.

Clark, J. Desmond. "African and Asian Perspectives on the Origin of Modern Humans." In *The Origin of Modern Humans and the Impact of Chronometric Dating*, edited by M. J. Aitken, C. B. Stringer, and P. A. Mellars, 148–178. Princeton, N.J.: Princeton University Press, 1993.

——. *Kalambo Falls Prehistoric Site III: The Earlier Cultures: Middle and Earlier Stone Age*. Cambridge: Cambridge University Press, 2001.

Clark, W. E. Le Gros. *History of the Primates*. Chicago: University of Chicago Press, 1959.

Clarke, David L. *Analytical Archaeology*. 2nd ed. New York: Columbia University Press, 1978.

Clottes, Jean. *Chauvet Cave: The Art of Earliest Times*. Salt Lake City: University of Utah Press, 2003.

Coe, Michael D. *The Maya*. 6th ed. New York: Thames and Hudson, 1999.

Collingwood, R. G. *The Idea of History*. Oxford: Clarendon Press, 1946.

Conard, Nicholas J. "A Female Figurine from the Basal Aurignacian of Hohle Fels Cave in Southwestern Germany." *Nature* 459 (2009): 248–252.

——. "Palaeolithic Ivory Sculptures from Southwestern Germany and the Origins of Figurative Art." *Nature* 426 (2003): 830–832.

Conard, Nicholas J., Maria Malina, and Susanne C. Münzel. "New Flutes Document the Earliest Musical Tradition in Southwestern Germany." *Nature* 460 (2009): 727–740.

Conroy, Glenn C. *Primate Evolution*. New York: Norton, 1990.

╳ Coolidge, Frederick L., and Thomas Wynn. *The Rise of Homo sapiens: The Evolution of Modern Thinking*. Malden, Mass.: Wiley-Blackwell, 2009.

——. "Working Memory, Its Executive Functions, and the Emergence of Modern Thinking." *Cambridge Archaeological Journal* 15 (2005): 5–26.

Coon, Carleton S. *The Origin of Races*. New York: Knopf, 1963.

Corballis, Michael C. *The Lopsided Ape: Evolution of the Generative Mind*. New York: Oxford University Press, 1991.

——. "Recursion as the Key to the Human Mind." In *From Mating to Mentality: Evaluating Evolutionary Psychology*, edited by Kim Sterelny and Julie Fitness, 155–171. New York: Psychology Press, 2003.

Cosmides, Leda, and John Tooby. "Origins of Domain Specificity: The Evolution of Functional Organization." In *Mapping the Mind: Domain Specificity in Cognition and Culture*, edited by Lawrence A. Hirschfield and Susan A. Gelman, 85–116. Cambridge: Cambridge University Press, 1994.

Cuomo, S. *Technology and Culture in Greek and Roman Antiquity*. Cambridge: Cambridge University Press, 2007.

Daniel, Glyn. *The First Civilizations: The Archaeology of Their Origins*. New York: Crowell, 1968.

Darwin, Charles. *The Descent of Man and Selection in Relation to Sex*. 2nd ed. London: John Murray, 1875.

——. *On the Origin of Species*. London: John Murray, 1859.

Davidson, Iain, and William Noble. "The Archaeology of Depiction: Traces of Depiction and Language." *Current Anthropology* 30 (1989): 125–155.

Dawkins, Richard. *The Ancestor's Tale: A Pilgrimage to the Dawn of Evolution*. Boston: Houghton Mifflin, 2004.

——. *The Selfish Gene*. New York: Oxford University Press, 1976.

Deacon, Terrence W. *The Symbolic Species: The Co-Evolution of Language and the Brain*. New York: Norton, 1997.

╳ de Beaune, Sophie A., Frederick L. Coolidge, and Thomas Wynn, eds. *Cognitive Archaeology and Human Evolution*. Cambridge: Cambridge University Press, 2009.

de Camp, L. Sprague. *The Ancient Engineers*. New York: Ballantine Books, 1963.

Deetz, James. *In Small Things Forgotten: The Archaeology of American Life*. Garden City, N.Y.: Anchor Books, 1977.

——. *Invitation to Archaeology*. Garden City, N.Y.: Natural History Press, 1967.

De Gusta, David, W. Henry Gilbert, and Scott P. Turner. "Hypoglossal Canal Size and Hominid Speech." *Proceedings of the National Academy of Sciences* 96 (1999): 1800–1804.

de la Torre, Ignacio, Rafael Mora, Manuel Domínguez-Rodrigo, Luis de Luque, and Luis Alcalá. "The Oldowan Industry of Peninj and Its Bearing on the Reconstruction of the Technological Skills of Lower Pleistocene Hominids." *Journal of Human Evolution* 44 (2003): 203–224.

DeMarrais, Elizabeth, Chris Gosden, and Colin Renfrew, eds. *Rethinking Materiality: The Engagement of Mind with the Material World*. Cambridge: McDonald Institute for Archaeological Research, 2004.

Dennett, Daniel C. *Brainstorms: Philosophical Essays on Mind and Psychology*. Cambridge, Mass.: MIT Press, 1978.

——. *Consciousness Explained*. Boston: Little, Brown, 1991.

d'Errico, Francesco. "The Invisible Frontier: A Multiple Species Model for the Origin of Behavioral Modernity." *Evolutionary Anthropology* 12 (2003): 188–202.

——. "Palaeolithic Origins of Artificial Memory Systems: An Evolutionary Perspective." In *Cognition and Material Culture: The Archaeology of Symbolic Storage*, edited by Colin Renfrew and Chris Scarre, 19–50. Cambridge: McDonald Institute for Archaeological Research, 1998.

d'Errico, Francesco, Christopher Henshilwood, Graeme Lawson, Marion Vanhaeren, Anne-Marie Tillier, Marie Soressi, Frédérique Bresson, et al. "Archaeological Evidence for the Emergence of Language, Symbolism, and Music—An Alternative Multidisciplinary Perspective." *Journal of World Prehistory* 17 (2003): 1–70.

d'Errico, Francesco, and April Nowell. "A New Look at the Berekhat Ram Figurine: Implications for the Origins of Symbolism." *Cambridge Archaeological Journal* 10 (2000): 123–167.

Descartes, René. *Philosophical Works*. 2 vols. Translated by Elizabeth S. Haldane and G. R. T. Ross. New York: Dover, 1955.

Dolukhanov, Pavel, Dmitry Sokoloff, and Anvar Shukurov. "Radiocarbon Chronology of Upper Palaeolithic Sites in Eastern Europe at Improved Resolution." *Journal of Archaeological Science* 28 (2001): 699–712.

Domínguez-Rodrigo, Manuel, and Travis Rayne Pickering. "Early Hominid Hunting and Scavenging: A Zooarchaeological Review." *Evolutionary Anthropology* 12 (2003): 275–282.

Donald, Merlin. *Origins of the Modern Mind: Three Stages in the Evolution of Culture and Cognition*. Cambridge, Mass.: Harvard University Press, 1991.

Dretske, Fred. *Explaining Behavior: Reasons in a World of Causes*. Cambridge, Mass.: MIT Press, 1988.

Drexler, K. Eric. *Engines of Creation: The Coming Era of Nanotechnology*. New York: Anchor Books, 1986.

Dunbar, Robin. *Grooming, Gossip, and the Evolution of Language*. Cambridge, Mass.: Harvard University Press, 1996.

——. "The Social Brain Hypothesis." *Evolutionary Anthropology* 6 (1998): 178–190.

Dyson, George B. *Darwin Among the Machines: The Evolution of Global Intelligence*. Reading, Mass.: Addison-Wesley, 1997.

Edelman, Gerald M. *Wider Than the Sky: The Phenomenal Gift of Consciousness*. New Haven, Conn.: Yale University Press, 2004.

Edelman, Gerald M., and Giulio Tononi. *A Universe of Consciousness: How Matter Becomes Imagination*. New York: Basic Books, 2000.

Edwards, I. E. S. *The Pyramids of Egypt*. Harmondsworth: Penguin, 1961.

Efimenko, P. P. *Pervobytnoe obshchestvo*. 2nd ed. Leningrad: USSR Academy of Sciences, 1938.

Eldredge, Niles. *Macro-Evolutionary Dynamics: Species, Niches, and Adaptive Peaks*. New York: McGraw-Hill, 1989.

——. *Time Frames: The Evolution of Punctuated Equilibria*. Princeton, N.J.: Princeton University Press, 1985.

Elvin, Mark. *The Pattern of the Chinese Past*. Stanford, Calif.: Stanford University Press, 1973.

Emmorey, Karen, Sonya Mehta, and Thomas J. Grabowski. "The Neural Correlates of Sign Versus Word Production." *NeuroImage* 36 (2007): 202–208.

Enard, Wolfgang, Molly Przeworski, Simon E. Fisher, Cecelia S. L. Lai, Victor Wiebe, Takashi Kitano, Anthony P. Monaco, and Svante Paabo. "Molecular Evolution of FOXP2, a Gene Involved in Speech and Language." *Nature* 418 (2002): 869–872.

✗ Fagan, Brian. *Floods, Famines, and Emperors: El Niño and the Fate of Civilizations.* New York: Basic Books, 1999.

——. *The Journey from Eden: The Peopling of Our World.* London: Thames and Hudson, 1990.

✗ ——. *The Long Summer: How Climate Changed Civilization.* New York: Basic Books, 2004.

Fedele, Francesco G., Biagio Giaccio, and Irka Hajdas. "Timescales and Cultural Processes at 40,000 BP in the Light of the Campanian Ignimbrite Eruption, Western Eurasia." *Journal of Human Evolution* 55 (2008): 834–857.

Fine, Cordelia, ed. *The Britannica Guide to the Brain.* London: Constable and Robinson, 2008.

Finley, M. I. *The Ancient Economy.* 2nd ed., updated. Berkeley: University of California Press, 1999.

Firth, Raymond. *Symbols: Public and Private.* Ithaca, N.Y.: Cornell University Press, 1973.

Fitch, W. Tecumseh. "The Evolution of Speech: A Comparative Review." *Trends in Cognitive Sciences* 4 (2000): 258–267.

Fitch, W. Tecumseh, and Marc D. Hauser. "Computational Constraints on Syntactic Processing in a Nonhuman Primate." *Science* 303 (2004): 377–380.

Fodor, Jerry A. *The Mind Doesn't Work That Way: Scope and Limits of Computational Psychology.* Cambridge, Mass.: MIT Press, 2000.

——. *The Modularity of Mind.* Cambridge, Mass.: MIT Press, 1983.

Foley, Robert, and Marta Mirazon Lahr. "On Stony Ground: Lithic Technology, Human Evolution, and the Emergence of Culture." *Evolutionary Anthropology* 12 (2003): 109–122.

Freeman, Charles. *The Greek Achievement: The Foundation of the Western World.* New York: Viking Press, 1999.

Frisch, Karl von. *The Dance Language and Orientation of Bees.* Translated by Leigh E. Chadwick. Cambridge, Mass.: Harvard University Press, 1967.

Galik, K., B. Senut, M. Pickford, D. Gommery, J. Treil, A. J. Kuperavage, and R. B. Eckhardt. "External and Internal Morphology of the BAR 1002'00 *Orrorin tugenensis* Femur." *Science* 305 (2004): 1450–1453.

✗ Gamble, Clive. *Origins and Revolutions: Human Identity in Earliest Prehistory.* Cambridge: Cambridge University Press, 2007.

——. *The Palaeolithic Settlement of Europe.* Cambridge: Cambridge University Press, 1986.

——. *The Palaeolithic Societies of Europe.* Cambridge: Cambridge University Press, 1999.

——. *Timewalkers: The Prehistory of Global Colonization.* Cambridge, Mass.: Harvard University Press, 1994.

Gans, C., and G. Northcutt. "Neural Crest and the Origin of the Vertebrates: A New Head." *Science* 220 (1983): 268–273.

Gathercole, Peter. "Childe's Revolutions." In *Archaeology: The Key Concepts,* edited by Colin Renfrew and Paul Bahn, 35–41. Abingdon: Routledge, 2005.

Germonpré, Mietje, Mikhail V. Sablin, Rhiannon E. Stevens, Robert E. M. Hedges, Michael Hofreiter, Mathias Stiller, and Viviane R. Després. "Fossil Dogs and Wolves from Palaeolithic Sites in Belgium, the Ukraine, and Russia: Osteometry, Ancient DNA, and Stable Isotopes." *Journal of Archaeological Science* 36 (2009): 473–490.

Gimpel, Jean. *The Medieval Machine: The Industrial Revolution of the Middle Ages.* New York: Holt, 1976.

Goebel, Ted. "The Pleistocene Colonization of Siberia and Peopling of the Americas: An Ecological Approach." *Evolutionary Anthropology* 8 (1999): 208–227.

Goldberg, Elkhonon. *The New Executive Brain: Frontal Lobes in a Complex World.* Oxford: Oxford University Press, 2009.

Golovanova, Liubov V., Vladimir B. Doronichev, Naomi E. Cleghorn, Marianna A. Kulkova, Tatiana V. Sapelko, and M. Steven Shackley. "Significance of Ecological Factors in the Middle to Upper Paleolithic Transition." *Current Anthropology* 51 (2010): 655–691.

Goodall, Jane. *In the Shadow of Man.* London: Collins, 1971.

Gosden, Christopher. *Social Being and Time.* Oxford: Blackwell, 1994.

Gowlett, John A. J. "The Elements of Design Form in Acheulian Bifaces: Modes, Modalities, Rules and Language." In *Axe Age: Acheulian Tool-making from Quarry to Discard*, edited by Naama Goren-Inbar and Gonen Sharon, 203–221. London: Equinox, 2006.

Graf, Kelly E. "The Good, the Bad, and the Ugly: Evaluating the Radiocarbon Chronology of the Middle and Late Upper Paleolithic in the Enisei River Valley, South-Central Siberia." *Journal of Archaeological Science* 36 (2009): 694–707.

Green, Richard E., Johannes Krause, Adrian W. Briggs, Tomislav Maricic, Udo Stenzel, Martin Kircher, Nick Patterson, et al. "A Draft Sequence of the Neandertal Genome." *Science* 328 (2010): 710–722

Green, Sally. *Prehistorian: A Biography of V. Gordon Childe.* Bradford-on-Avon: Moonraker Press, 1981.

Greene, Kevin. "Technological Innovation and Economic Progress in the Ancient World: M. I. Finley Reconsidered." *Economic History Review*, n.s., 53 (2000): 29–59.

——. "V. Gordon Childe and the Vocabulary of Revolutionary Change." *Antiquity* 73 (1999): 97–109.

Greenfield, Patricia M. "Language, Tools, and Brain: The Ontogeny and Phylogeny of Hierarchically Organized Sequential Behavior." *Behavioral and Brain Sciences* 14 (1991): 531–595.

Hagoort, Peter, Lea Hald, Marcel Bastiaansen, and Karl Magnus Petersson. "Integration of Word Meaning and World Knowledge in Language Comprehension." *Science* 304 (2004): 438–441.

Haidle, Miriam Noël. "How to Think a Spear." In *Cognitive Archaeology and Human Evolution*, edited by Sophie A. de Beaune, Frederick L. Coolidge, and Thomas Wynn, 57–73. Cambridge: Cambridge University Press, 2009.

Hannah, Robert. "Timekeeping." In *The Oxford Handbook of Engineering and Technology in the Classical World*, edited by John Peter Oleson, 740–758. Oxford: Oxford University Press, 2008.

Hapgood, Philip J., and Natalie R. Franklin. "The Revolution That Didn't Arrive: A Review of Pleistocene Sahul." *Journal of Human Evolution* 55 (2008): 187–222.

Hardy, Bruce L., Marvin Kay, Anthony E. Marks, and Katherine Monigal. "Stone Tool Function at the Paleolithic Sites of Starosele and Buran Kaya III, Crimea: Behavioral Implications." *Proceedings of the National Academy of Sciences* 98 (2001): 10972–10977.

Harris, Marvin. *The Rise of Anthropological Theory: A History of Theories of Culture.* New York: Crowell, 1968.

Hauser, Marc D., Noam Chomsky, and W. Tecumseh Fitch. "The Faculty of Language: What Is It, Who Has It, and How Did It Evolve?" *Science* 298 (2002): 1569–1579.

Headrick, Daniel R. *Technology: A World History.* Oxford: Oxford University Press, 2009.

Hegel, Georg Wilhelm Friedrich. *Philosophy of History.* Translated by J. Sibree. New York: Barnes & Noble, 2004.

Heidegger, Martin. *The Question Concerning Technology and Other Essays.* Translated by William Lovitt. New York: Harper & Row, 1977.

Henshilwood, Christopher. "Fully Symbolic *Sapiens* Behaviour: Innovation in the Middle Stone Age at Blombos Cave, South Africa." In *Rethinking the Human Revolution: New Behavioural and Biological Perspectives on the Origin and Dispersal of Modern Humans,* edited by Paul Mellars, Katie Boyle, Ofer Bar-Yosef, and Chris Stringer, 123–132. Cambridge: McDonald Institute for Archaeological Research, 2007.

Henshilwood, Christopher, and Curtis W. Marean. "The Origin of Modern Human Behavior: Critique of the Models and Their Test Implications." *Current Anthropology* 44 (2003): 627–651.

Hewes, Gordon W. "A History of Speculation on the Relation Between Tools and Language." In *Tools, Language and Cognition in Human Evolution,* edited by Kathleen R. Gibson and Tim Ingold, 20–31. Cambridge: Cambridge University Press, 1993.

Hillman, Gordon C. "Late Pleistocene Changes in Wild Plant Foods Available to Hunter-Gatherers of the Northern Fertile Crescent: Possible Preludes to Cereal Cultivation." In *The Origins and Spread of Agriculture and Pastoralism in Eurasia,* edited by David R. Harris, 159–203. Washington, D.C.: Smithsonian Institution Press, 1996.

Hillman, Gordon C., and M. Stuart Davies. "Measured Domestication Rates in Wild Wheat and Barley Under Primitive Cultivation and Their Archaeological Implications." *Journal of World Prehistory* 4 (1990): 157–222.

Hirshfeld, Alan. *Eureka Man: The Life and Legacy of Archimedes.* New York: Walker, 2009.

Hiscock, Peter. *Archaeology of Ancient Australia.* London: Routledge, 2008.

Hobbes, Thomas. *Leviathan, or the Matter, Forme and Power of a Common-wealth Ecclesiasticall and Civill.* London: Penguin, 1968.

Hodder, Ian. *Reading the Past: Current Approaches to Interpretation in Archaeology.* 2nd ed. Cambridge: Cambridge University Press, 1991.

——. "Theoretical Archaeology: A Reactionary View." In *Symbolic and Structural Archaeology,* edited by Ian Hodder, 12–13. Cambridge: Cambridge University Press, 1981.

Hoffecker, John F. *Desolate Landscapes: Ice-Age Settlement of Eastern Europe.* New Brunswick, N.J.: Rutgers University Press, 2002.

——. "The Eastern Gravettian 'Kostenki Culture' as an Arctic Adaptation." *Anthropological Papers of the University of Alaska,* n.s., 2 (2002): 115–136.

——. "Innovation and Technological Knowledge in the Upper Paleolithic of Northern Eurasia." *Evolutionary Anthropology* 14 (2005): 186–198.

——. *A Prehistory of the North: Human Settlement of the Higher Latitudes.* New Brunswick, N.J.: Rutgers University Press, 2005.

——. "Representation and Recursion in the Archaeological Record." *Journal of Archaeological Method and Theory* 14 (2007): 370–375.

——. "The Spread of Modern Humans in Europe." *Proceedings of the National Academy of Sciences* 106 (2009): 16040–16045.

Hoffecker, John F., and Scott A. Elias. *Human Ecology of Beringia.* New York: Columbia University Press, 2007.

Hoffecker, John F., Vance T. Holliday, Mikhail V. Anikovich, Andrei A. Sinitsyn, V. V. Popov, S. N. Lisitsyn, G. M. Levkovskaya, G. A. Pospelova, S. L. Forman, and B. Giaccio. "From the Bay of Naples to the River Don: The Campanian Ignimbrite Eruption and the Middle to Upper Paleolithic Transition in Eastern Europe." *Journal of Human Evolution* 55 (2008): 858–870.

Hoffecker, John F., I. E. Kuz'mina, E. V. Syromyatnikova, Mikhail V. Anikovich, Andrei A. Sinitsyn, V. V. Popov, and Vance T. Holliday. "Evidence for Kill-Butchery Events of Early Upper Paleolithic Age at Kostenki, Russia." *Journal of Archaeological Science* 37 (2010): 1073–1089.

Hofstadter, Douglas R. *Gödel, Escher, Bach: An Eternal Golden Braid.* New York: Basic Books, 1979.

Hole, Frank. "Environmental Instabilities and Urban Origins." In *Chiefdoms and Early States in the Near East: The Organizational Dynamics of Complexity,* edited by Gil Stein and Mitchell S. Rothman, 121–143. Madison, Wis.: Prehistory Press, 1994.

Hölldobler, Bert, and E. O. Wilson, *The Super-Organism: The Beauty, Elegance, and Strangeness of Insect Societies.* New York: Norton, 2009.

Holliday, Trenton W. "Brachial and Crural Indices of European Late Upper Paleolithic and Mesolithic Humans." *Journal of Human Evolution* 36 (1999): 549–566.

——. "Postcranial Evidence of Cold Adaptation in European Neandertals." *American Journal of Physical Anthropology* 104 (1997): 245–258.

Holliday, Vance T., John F. Hoffecker, Paul Goldberg, Richard I. Macphail, Steven L. Forman, Mikhail Anikovich, and Andrei Sinitsyn. "Geoarchaeology of the Kostenki-Borshchevo Sites, Don River Valley, Russia." *Geoarchaeology: An International Journal* 22 (2007): 181–228.

Holloway, Ralph L. "Brief Communication: How Much Larger Is the Relative Volume of Area 10 of the Prefrontal Cortex in Humans?" *American Journal of Physical Anthropology* 118 (2002): 399–401.

——. "Culture: A *Human* Domain." *Current Anthropology* 10 (1969): 395–412.

——. "The Poor Brain of *Homo sapiens neanderthalensis*: See What You Please . . ." In *Ancestors: The Hard Evidence,* edited by Eric Delson, 319–324. New York: Liss, 1985.

Holloway, Ralph L., Chet S. Sherwood, Patrick R. Hof, and James K. Rilling. "Evolution of the Brain in Humans—Paleoneurology." In *Encyclopedia of Neuroscience,* edited by Mark D. Binder, Nobutaka Hirokawa, and Uwe Windhorst, 1326–1334. New York: Springer, 2009.

Hughes, Thomas P. *Human-Built World: How to Think About Technology and Culture.* Chicago: University of Chicago Press, 2004.

Isaac, Glynn Ll. *Olorgesailie: Archeological Studies of a Middle Pleistocene Lake Basin in Kenya*. Chicago: University of Chicago Press, 1977.

Jarvis, J. U. M. "Eusociality in a Mammal: Cooperative Breeding in Naked Mole-Rat Colonies." *Science* 212 (1981): 571–573.

Jelinek, Arthur J. "The Lower Paleolithic: Current Evidence and Interpretation." *Annual Review of Anthropology* 6 (1977): 11–32.

Jones, Martin. "Moving North: Archaeobotanical Evidence for Plant Diet in Middle and Upper Paleolithic Europe." In *The Evolution of Hominin Diets: Integrating Approaches to the Study of Palaeolithic Subsistence*, edited by Jean-Jacques Hublin and Michael P. Richards, 171–180. Dordrecht: Springer, 2009.

Jones, Rhys. "East of Wallace's Line: Issues and Problems in the Colonisation of the Australian Continent." In *The Human Revolution: Behavioural and Biological Perspectives on the Origins of Modern Humans*, edited by Paul Mellars and Chris Stringer, 743–782. Princeton, N.J.: Princeton University Press, 1989.

Kant, Immanuel. *On History*. Edited by Lewis White Beck. Translated by Lewis White Beck, Robert E. Anchor, and Emil L. Fackenheim. New York: Macmillan, 1963.

Kasarda, John D. "The Structural Implications of Social System Size: A Three-Level Analysis." *American Sociological Review* 39 (1974): 19–28.

Kay, Richard F., Matt Cartmill, and Michelle Balow. "The Hypoglossal Canal and the Origin of Human Vocal Behavior." *Proceedings of the National Academy of Sciences* 95 (1988): 5417–5419.

Keally, Charles T., Yasuhiro Taniguchi, and Yaroslav V. Kuzmin. "Understanding the Beginnings of Pottery Technology in Japan and Neighboring East Asia." *Review of Archaeology* 24 (2003): 3–14.

Keeley, Lawrence H. "Microwear Analysis of Lithics." In *The Lower Paleolithic Site at Hoxne, England*, edited by Ronald Singer, Bruce G. Gladfelter, and John J. Wymer, 129–138. Chicago: University of Chicago Press, 1993.

Keeley, Lawrence H., and Nicholas Toth. "Microwear Polishes on Early Stone Tools from Koobi Fora, Kenya." *Nature* 293 (1981): 464–466.

Kelly, Robert L. *The Foraging Spectrum: Diversity in Hunter-Gatherer Lifeways*. Washington, D.C.: Smithsonian Institution Press, 1995.

Kemp, Barry J. *Ancient Egypt: Anatomy of a Civilization*. London: Routledge, 1989.

Kilbride-Jones, Howard. "From Mr Howard Kilbride-Jones, 6 April 1992." In *The Archaeology of V. Gordon Childe*, edited by David R. Harris, 135–139. Chicago: University of Chicago Press, 1994.

Kim, Jaegwon. *Philosophy of Mind*. 2nd ed. Boulder, Colo.: Westview Press, 2006.

Kingdon, Jonathan. *Lowly Origin: Where, When, and Why Our Ancestors First Stood Up*. Princeton, N.J.: Princeton University Press, 2003.

Kirch, Patrick Vinton. *On the Road of the Winds: An Archaeological History of the Pacific Islands Before European Contact*. Berkeley: University of California Press, 2000.

Klein, Richard G. *The Human Career: Human Biological and Cultural Origins*. 2nd ed. Chicago: University of Chicago Press, 1999.

——. *The Human Career: Human Biological and Cultural Origins*. 3rd ed. Chicago: University of Chicago Press, 2009.

——. *Man and Culture in the Late Pleistocene: A Case Study*. San Francisco: Chandler, 1969.

Knecht, Heidi. "Splits and Wedges: The Techniques and Technology of Early Aurignacian Antler Working." In *Before Lascaux: The Complex Record of the Early Upper Paleolithic*, edited by Heidi Knecht, Anne Pike-Tay, and Randall White, 137–162. Boca
Raton, Fla.: CRC Press, 1993.

Kozlowski, Janusz K. "The Gravettian in Central and Eastern Europe." In *Advances in
World Archaeology*, edited by Fred Wendorf and Angela E. Close, 5:131–200. Orlando,
Fla.: Academic Press, 1986.

Kramer, Samuel Noah. *The Sumerians: Their History, Culture, and Character*. Chicago: University of Chicago Press, 1963.

Kuhn, Steven L., and Amilcare Bietti. "The Late Middle and Early Upper Paleolithic in
Italy." In *The Geography of Neandertals and Modern Humans in Europe and the Greater
Mediterranean*, edited by Ofer Bar-Yosef and David Pilbeam, 49–76. Cambridge,
Mass.: Peabody Museum of Archaeology and Ethnology, 2000.

Kuhn, Steven L., Mary C. Stiner, Erksin Güleç, Ismail Özer, Hakan Yılmaz, Ismail
Baykara, Ayşen Açıkkol, et al. "The Early Upper Paleolithic Occupations at Üçağizli
Cave (Hatay, Turkey)." *Journal of Human Evolution* 56 (2009): 87–113.

Kurzweil, Ray. *The Age of Spiritual Machines: When Computers Exceed Human Intelligence*.
New York: Viking Press, 1999.

———. *The Singularity Is Near: When Humans Transcend Biology*. New York: Viking Press, 2005.

Kyriacou, Andreas. "Innovation and Creativity: A Neuropsychological Perspective." In
Cognitive Archaeology and Human Evolution, edited by Sophie A. de Beaune, Frederick L. Coolidge, and Thomas Wynn, 15–24. Cambridge: Cambridge University Press,
2009.

Laitman, Jeffrey. "The Anatomy of Human Speech." *Natural History*, August 1984, 20–27.

Landels, J. G. *Engineering in the Ancient World*. Berkeley: University of California Press,
1978.

Landes, David S. *Revolution in Time: Clocks and the Making of the Modern World*. Cambridge, Mass.: Harvard University Press, 1983.

———. *The Wealth and Poverty of Nations: Why Some Are So Rich and Some So Poor*. New York:
Norton, 1999.

Langdon, John H. *The Human Strategy: An Evolutionary Perspective on Human Anatomy*.
New York: Oxford University Press, 2005.

Lavan, Luke. "Explaining Technological Change: Innovation, Stagnation, Recession and
Replacement." In *Technology in Transition, A.D. 300–650*, edited by Luke Lavan, Enrico
Zanini, and Alexander Sarantis, xv–xl. Leiden: Brill, 2007.

Leach, Edmund. *Claude Lévi-Strauss*. New York: Viking Press, 1970.

Leakey, Mary D. *Olduvai Gorge*. Vol. 3, *Excavations in Beds I and II, 1960–1963*. Cambridge: Cambridge University Press, 1971.

Leakey, Meave G., Craig S. Feibel, Ian MacDougall, and Alan Walker. "New Four-
Million-Year-Old Hominid Species from Kanapoi and Allia Bay, Kenya." *Nature* 376
(1995): 565–571.

LeDoux, Joseph. *Synaptic Self: How Our Brains Become Who We Are*. New York: Penguin,
2002.

Leroi-Gourhan, André. *Le Geste et la parole*. Vol. 1, *Technique et langage*. Paris: Albin
Michel, 1964.

——. *Le Geste et la parole.* Vol. 2, *La Mémoire et les rythmes.* Paris: Albin Michel, 1965.

——. *Gesture and Speech.* Translated by Anna Bostock Berger. Cambridge, Mass.: MIT Press, 1993.

Lévi-Strauss, Claude. *The Raw and the Cooked: Introduction to a Science of Mythology:* Vol. 1. Translated by John Weightman and Doreen Weightman. New York: Harper & Row, 1969.

——. *The Savage Mind.* Chicago: University of Chicago Press, 1966.

——. *Structural Anthropology.* Translated by Claire Jacobson and Brooke Grundfest Scho-epf. New York: Basic Books, 1963.

Levitin, Daniel J. *This Is Your Brain on Music: The Science of a Human Obsession.* New York: Penguin, 2006.

Lewin, Roger, and Robert A. Foley. *Principles of Human Evolution.* Malden, Mass.: Black-well, 2004.

Lewontin, Richard C. *The Genetic Basis of Evolutionary Change.* New York: Columbia University Press, 1974.

Lieberman, Daniel E. "Speculations About the Selective Basis for Modern Human Craniofacial Form." *Evolutionary Anthropology* 17 (2008): 55–68.

Lieberman, Philip. *The Biology and Evolution of Language.* Cambridge, Mass.: Harvard University Press, 1984.

——. *Eve Spoke: Human Language and Human Evolution.* New York: Norton, 1998.

Lindberg, David C. *The Beginnings of Western Science: The European Scientific Tradition in Philosophical, Religious, and Institutional Context, 600 B.C. to A.D. 1450.* Chicago: University of Chicago Press, 1992.

Lloyd, Geoffrey. *Greek Science After Aristotle.* London: Chatto and Windus, 1973.

Lombard, Marlize. "Evidence of Hunting and Hafting During the Middle Stone Age at Sibudu Cave, KwaZulu-Natal: A Multianalytical Approach." *Journal of Human Evolution* 48 (2005): 279–300.

——. "The Gripping Nature of Ochre: The Association of Ochre with Howiesons Poort Adhesives and Later Stone Age Mastics from South Africa." *Journal of Human Evolution* 53 (2007): 406–419.

Lordkipanidze, David, Tea Jashashvili, Abesalom Vekua, Marcia S. Ponce de León, Christoph P. E. Zollikofer, G. Philip Rightmire, Herman Pontzer, et al. "Postcranial Evidence from Early *Homo* from Dmanisi, Georgia." *Nature* 449 (2007): 305–310.

Lourandos, Harry. *Continent of Hunter-Gatherers: New Perspectives in Australian Prehistory.* Cambridge: Cambridge University Press, 1997.

Lucas, A., and J. R. Harris. *Ancient Egyptian Materials and Industries.* 4th ed. London: Arnold, 1962.

Lycett, Stephen J. "Are Victoria West Cores 'Proto-Levallois'? A Phylogenetic Assessment." *Journal of Human Evolution* 56 (2009): 175–191.

Maisels, Charles Keith. *Early Civilizations of the Old World: The Formative Histories of Egypt, the Levant, Mesopotamia, India, and China.* London: Routledge, 1999.

Mania, Dietrich, and Ursula Mania. "Deliberate Engravings on Bone Artefacts of *Homo erectus.*" *Rock Art Research* 5 (1988): 91–107.

Mania, Ursula "The Utilisation of Large Mammal Bones in Bilzingsleben—A Special Variant of Middle Pleistocene Man's Relationship to His Environment." *ERAUL* 62 (1995): 239–246.

Marks, Anthony E., and C. Reid Ferring. "The Early Upper Paleolithic of the Levant." In *The Early Upper Paleolithic: Evidence from Europe and the Near East*, edited by John F. Hoffecker and Cornelia A. Wolf, 43-72. Oxford: British Archaeological Reports, 1988.

Marr, David. *Vision: A Computational Investigation into the Human Representation and Processing of Visual Information*. San Francisco: Freeman, 1982.

Marshack, Alexander. *The Roots of Civilization: The Cognitive Beginnings of Man's First Art, Symbol and Notation*. New York: McGraw-Hill, 1972.

Martin, Robert. *Primate Origins and Evolution: A Phylogenetic Reconstruction*. Princeton, N.J.: Princeton University Press, 1990.

Marx, Karl, and Frederick Engels. *The German Ideology*. Part One. Edited by C. J. Arthur. New York: International Publishers, 1947.

Marzke, Mary W. "Joint Function and Grips of the *Australopithecus afarensis* Hand, with Special Reference to the Region of the Capitate." *Journal of Human Evolution* 12 (1983): 197-211.

Maslin, Keith T. *An Introduction to the Philosophy of Mind*. 2nd ed. Cambridge: Polity, 2007.

Mayr, Ernst. *The Growth of Biological Thought: Diversity, Evolution, and Inheritance*. Cambridge, Mass.: Harvard University Press, 1982.

Mayr, Ernst, and William B. Provine, eds. The *Evolutionary Synthesis: Perspectives on the Unification of Biology*. Cambridge, Mass.: Harvard University Press, 1980.

McBrearty, Sally, and Alison S. Brooks. "The Revolution That Wasn't: A New Interpretation of the Origin of Modern Human Behavior." *Journal of Human Evolution* 39 (2000): 453-563.

McGhee, Robert. *Ancient People of the Arctic*. Vancouver: University of British Columbia Press, 1996.

McGrew, William. *The Cultured Chimpanzee: Reflections on Cultural Primatology*. Cambridge: Cambridge University Press, 2004.

McLeish, John. *The Story of Numbers: How Mathematics Shaped Civilization*. New York: Fawcett Columbine, 1991.

McNeill, J. R., and William H. McNeill. *The Human Web: A Bird's-Eye View of World History*. New York: Norton, 2003.

McNeill, William H. *Plagues and Peoples*. Garden City, N.Y.: Anchor Books, 1976.

——. *The Pursuit of Power: Technology, Armed Force, and Society Since* A.D. *1000*. Chicago: University of Chicago Press, 1982.

McPherron, Shannon P. "What Typology Can Tell Us About Acheulian Handaxe Production." In *Axe Age: Acheulian Tool-making from Quarry to Discard*, edited by Naama Goren-Inbar and Gonen Sharon, 267-285. London: Equinox, 2006.

Mednick, Sarnoff A. "The Associative Basis of the Creative Process." *Psychological Review* (1962): 220-232.

Mellars, Paul. "Going East: New Genetic and Archaeological Perspectives on the Modern Human Colonization of Eurasia." *Science* 313 (2006): 796-800.

——. *The Neanderthal Legacy: An Archaeological Perspective from Western Europe*. Princeton, N.J.: Princeton University Press, 1996.

Minsky, Marvin. *The Society of Mind*. New York: Simon and Schuster, 1986.

Mitcham, Carl. *Thinking Through Technology: The Path Between Engineering and Philosophy*. Chicago: University of Chicago Press, 1994.

Mitcham, Carl, and Robert Mackey, eds. *Philosophy and Technology: Readings in the Philosophical Problems of Technology*. New York: Free Press, 1983.

Mithen, Steven. *The Prehistory of the Mind: The Cognitive Origins of Art, Religion and Science*. London: Thames and Hudson, 1996.

Mokyr, Joel. *The Lever of Riches: Technological Creativity and Economic Progress*. New York: Oxford University Press, 1990.

Monahan, Christoph M. "New Zooarchaeological Data from Bed II, Olduvai Gorge, Tanzania: Implications for Hominid Behavior in the Early Pleistocene." *Journal of Human Evolution* 31 (1996): 93–128.

Moore, A. M. T., G. C. Hillman, and A. J. Legge. *Village on the Euphrates: From Foraging to Farming at Abu Hureyra*. London: Oxford University Press, 2000.

Moorehead, Alan. *The Fatal Impact: An Account of the Invasion of the South Pacific, 1767–1840*. New York: Harper & Row, 1966.

More, Thomas. *Utopia*. Translated by Paul Turner. London: Penguin, 1965.

Mountcastle, Vernon B. *Perceptual Neuroscience: The Cerebral Cortex*. Cambridge, Mass.: Harvard University Press, 1998.

——. *The Sensory Hand: Neural Mechanisms of Somatic Sensation*. Cambridge, Mass.: Harvard University Press, 2005.

Moyà-Solà, Salvador, and Meike Köhler. "A *Dryopithecus* Skeleton and the Origin of Great-Ape Locomotion." *Nature* 379 (1996): 156–159.

Mumford, Lewis. *The Myth of the Machine: Technics and Human Development*. New York: Harcourt, Brace & World, 1967.

——. *Technics and Civilization*. New York: Harcourt, Brace, 1934.

Nabokov, Vladimir. *Lolita*. New York: Putnam, 1955.

——. *Speak, Memory: An Autobiography Revisited*. New York: Vintage, 1967.

Napier, John. "The Evolution of the Hand." *Scientific American* 207 (1962): 56–62.

——. "Fossil Hand Bones from Olduvai Gorge." *Nature* 196 (1962): 409–411.

——. *Hands*. New York: Pantheon Books, 1980.

——. *Hands*. Revised by Russell H. Tuttle. Princeton, N.J.: Princeton University Press, 1993.

Needham, Joseph. *The Development of Iron and Steel Technology in China*. Cambridge: Heffer, 1964.

——. *Science and Civilization in China*. Vol. 4, part 2, *Mechanical Engineering*. Cambridge: Cambridge University Press, 1965.

——. *Science in Traditional China*. Cambridge, Mass.: Harvard University Press, 1981.

——. *The Shorter Science and Civilization in China*. Vol. 3. Cambridge: Cambridge University Press, 1986.

Nitecki, Matthew H., ed. *Evolutionary Innovations*. Chicago: University of Chicago Press, 1990.

Noble, William, and Iain Davidson. "The Evolutionary Emergence of Modern Human Behaviour: Language and Its Archaeology." *Man* 26 (1991): 223–253.

——. *Human Evolution, Language and Mind: A Psychological and Archaeological Inquiry*. Cambridge: Cambridge University Press, 1996.

Oleson, John Peter, ed. *The Oxford Handbook of Engineering and Technology in the Classical World*. Oxford: Oxford University Press, 2008.

Osborn, Henry Fairfield. *Men of the Old Stone Age: Their Environment, Life, and Art.* New York: Scribner, 1915.

Oswalt, Wendell H. *An Anthropological Analysis of Food-Getting Technology.* New York: Wiley, 1976.

Pacey, Arnold. *The Maze of Ingenuity: Ideas and Idealism in the Development of Technology.* 2nd ed. Cambridge, Mass.: MIT Press, 1992.

———. *Technology in World Civilization: A Thousand-Year History.* Cambridge, Mass.: MIT Press, 1990.

Parker, G. H. *The Elementary Nervous System.* Philadelphia: Lippincott, 1919.

Pearson, Osbjorn M. "Statistical and Biological Definitions of 'Anatomically Modern' Humans: Suggestions for a Unified Approach to Modern Morphology." *Evolutionary Anthropology* 17 (2008): 38–48.

Pelegrin, Jacques. "Cognition and the Emergence of Language: A Contribution from Lithic Technology." In *Cognitive Archaeology and Human Evolution,* edited by Sophie A. de Beaune, Frederick L. Coolidge, and Thomas Wynn, 95–108. Cambridge: Cambridge University Press, 2009.

Perrot, Jean. "Le gisement natoufien de Mallaha (Eynan), Israel." *L'Anthropologie* 70 (1966): 437–484.

Petraglia, Michael D. "The Lower Palaeolithic of India and Its Bearing on the Asian Record." In *Early Human Behaviour in Global Context: The Rise and Diversity of the Lower Paleolithic Record,* edited by Michael D. Petraglia and Ravi Korisettar, 343–390. London: Routledge, 1998.

Peyrony, Denis. *Élements de préhistoire.* Ussel: Jacques Eyboulet, 1934.

Piaget, Jean. *Structuralism.* Translated by Chaninah Maschler. New York: Harper & Row, 1970.

Pidoplichko, I. G. *Mezhirichskie zhilishcha iz Kostei Mamonta.* Kiev: Naukova dumka, 1976.

———. *Pozdnepaleoliticheskie zhilishcha iz kostei mamonta na Ukraine.* Kiev: Naukova dumka, 1969.

Piggott, Stuart. *The Earliest Wheeled Transport: From the Atlantic Coast to the Caspian Sea.* London: Thames and Hudson, 1983.

Pinker, Steven. *How the Mind Works.* New York: Norton, 1997.

Pinker, Steven, and Jacques Mehler. *Connections and Symbols.* Cambridge, Mass.: MIT Press, 1988.

Piperno, Dolores R., Ehud Weiss, Irene Holst, and Dani Nadel. "Processing of Wild Cereal Grains in the Upper Palaeolithic Revealed by Starch Grain Analysis." *Nature* 430 (2005): 670–673.

Pitts, Michael, and Mark Roberts. *Fairweather Eden: Life in Britain Half a Million Years Ago as Revealed by the Excavations at Boxgrove.* London: Century, 1998.

Pool, Robert. *Beyond Engineering: How Society Shapes Technology.* New York: Oxford University Press, 1997.

Potts, Richard, and Pat Shipman. "Cutmarks Made by Stone Tools on Bones from Olduvai Gorge, Tanzania." *Nature* 291 (1981): 577–580.

Povinelli, Daniel J. *Folk Physics for Apes: The Chimpanzee's Theory of How the World Works.* Oxford: Oxford University Press, 2000.

Putnam, Hilary. *Representation and Reality.* Cambridge, Mass.: MIT Press, 1988.

Raichle, Marcus E., Julie A. Fiez, Tom O. Videen, Ann-Mary K. MacLeod, Jose V. Pardo, Peter T. Fox, and Steven E. Petersen. "Practice-related Changes in Human Brain Functional Anatomy During Nonmotor Learning." *Cerebral Cortex* 4 (1994): 8–26.

Rappaport, Roy A. *Ritual and Religion in the Making of Humanity.* Cambridge: Cambridge University Press, 1999.

Raynal, Jean-Paul, Fatima-Zohra Sbihi Alaoui, Denis Geraads, Lionel Magoga, and Abderrahim Mohib. "The Earliest Occupation of North-Africa: The Moroccan Perspective." *Quaternary International* 75 (2001): 65–75.

Renfrew, Colin. "Cognitive Archaeology." In *Archaeology: The Key Concepts*, edited by Colin Renfrew and Paul Bahn, 41–45. Abingdon: Routledge, 2005.

———. *Towards an Archaeology of Mind.* Cambridge: Cambridge University Press, 1982.

———. "What Is Cognitive Archaeology?" *Cambridge Archaeological Journal* 3 (1993): 247–270.

Renfrew, Colin, and Chris Scarre, eds. *Cognition and Material Culture: The Archaeology of Symbolic Storage.* Cambridge: McDonald Institute for Archaeological Research, 1998.

Richards, Michael P., Paul B. Pettitt, Mary C. Stiner, and Erik Trinkaus. "Stable Isotope Evidence for Increasing Dietary Breadth in the European Mid-Upper Paleolithic." *Proceedings of the National Academy of Sciences* 98 (2001): 6528–6532.

Rightmire, G. Philip. "Human Evolution in the Middle Pleistocene: The Role of *Homo heidelbergensis.*" *Evolutionary Anthropology* 6 (1998): 218–227.

———. "*Homo* in the Middle Pleistocene: Hypodigms, Variation, and Species Recognition." *Evolutionary Anthropology* 17 (2008): 8–21.

Roberts, Alice. *The Incredible Human Journey: The Story of How We Colonised the Planet.* London: Bloomsbury, 2009.

Roche, H., A. Delagnes, J.-P. Brugal, C. Feibel, M. Kibunija, V. Mourre, and P.-J. Tixier. "Early Hominid Stone Tool Production and Technical Skill 2.34 Myr Ago in West Turkana, Kenya." *Nature* 399 (1999): 57–60.

Rodman, Peter S., and Henry M. McHenry. "Bioenergetics and the Origin of Hominid Bipedalism." *American Journal of Physical Anthropology* 52 (1980): 103–106.

Roe, Derek. "British Lower and Middle Palaeolithic Handaxe Groups." *Proceedings of the Prehistoric Society* 34 (1968): 245–267.

Rogachev, A. N. "Mnogosloinye Stoyanki Kostenkovsko-Borshevskogo raiona na Donu i Problema Razvitiya Kul'tury v Epokhy Verkhnego Paleolita na Russkoi Ravnine." *Materialy i Issledovaniya po Arkheologii SSSR* 59 (1957): 9–134.

Rogachev, A. N., N. D. Praslov, M. V. Anikovich, V. I. Belyaeva, and T. N. Dmitrieva. "Kostenki 1 (Stoyanka Polyakova)." In *Paleolit Kostenkovsko-Borshchevskogo Raiona na Donu, 1879–1979*, edited by N. D. Praslov and A. N. Rogachev, 42–66. Leningrad: Nauka, 1982.

Rogachev, A. N., and A. A. Sinitsyn. "Kostenki 15 (Gorodtsovskaya Stoyanka)." In *Paleolit Kostenkovsko-Borshchevskogo Raiona na Donu, 1879–1979*, edited by N. D. Praslov and A. N. Rogachev, 162–171. Leningrad: Nauka, 1982.

Rolland, Nicholas. "Levallois Technique Emergence: Single or Multiple? A Review of the Euro-African Record." In *The Definition and Interpretation of Levallois Technology*, edited by Harold L Dibble and Ofer Bar-Yosef, 333–359. Madison, Wis.: Prehistory Press, 1995.

Roux, Georges. *Ancient Iraq.* 3rd ed. London: Penguin, 1992.

Rowe, James B., Ivan Toni, Oliver Josephs, Richard S. J. Frackowiak, and Richard E. Passingham. "The Prefrontal Cortex: Response Selection or Maintenance Within Working Memory?" *Science* 288 (2000): 1656–1560.

Russon, Anne E., and David R. Begun. "Evolutionary Origins of Great Ape Intelligence: An Integrated View." In *The Evolution of Thought: Evolutionary Origins of Great Ape Intelligence*, edited by Anne E. Russon and David R. Begun, 353–368. Cambridge: Cambridge University Press, 2004.

Ryle, Gilbert. *The Concept of Mind*. New York: Barnes & Noble, 1949.

Sablin, Mikhail V., and Gennady A. Khlopachev. "The Earliest Ice Age Dogs: Evidence from Eliseevichi I." *Current Anthropology* 43 (2002): 795–799.

Sackett, James R. "The Neuvic Group: Upper Paleolithic Open-Air Sites in the Perigord." In *Upper Pleistocene Prehistory of Western Eurasia*, edited by Harold L. Dibble and Anna Montet-White, 61–84. Philadelphia: University of Pennsylvania Museum, 1988.

Santonja, Manuel, and Paola Villa. "The Acheulean of Western Europe." In *Axe Age: Acheulian Tool-making from Quarry to Discard*, edited by Naama Goren-Inbar and Gonen Sharon, 429–478. London: Equinox, 2006.

Savolainen, Peter, Ya-ping Zhang, Jing Luo, Joakim Lundeberg, and Thomas Leitner. "Genetic Evidence for an East Asian Origin of Domestic Dogs." *Science* 298 (2002): 1610–1613.

Schick, Kathy, and Nicholas Toth. *Making Silent Stones Speak: Human Evolution and the Dawn of Technology*. New York: Simon and Schuster, 1993.

———. "An Overview of the Oldowan Industrial Complex: The Sites and the Nature of Their Evidence." In *The Oldowan: Case Studies into the Earliest Stone Age*, edited by Nicholas Toth and Kathy Schick. Gosport, Ind.: Stone Age Institute Press, 2006.

Schlanger, Nathan. "The Chaîne Opératoire." In *Archaeology: The Key Concepts*, edited by Colin Renfrew and Paul Bahn, 25–31. Abingdon: Routledge, 2005.

———. "Understanding Levallois: Lithic Technology and Cognitive Archaeology." *Cambridge Archaeological Journal* 6 (1996): 231–254.

Scott, Gary R., and Luis Gilbert. "The Oldest Hand-Axes in Europe." *Nature* 461 (2009): 82–85.

Scott, Katherine. "Two Hunting Episodes of Middle Palaeolithic Age at La Cotte de Saint-Brelade, Jersey (Channel Islands)." *World Archaeology* 12 (1980): 137–152.

Searle, John R. *The Mind: A Brief Introduction*. New York: Oxford University Press, 2004.

Seeley, Thomas D. *Honeybee Democracy*. Princeton, N.J.: Princeton University Press, 2010.

Semaw, Sileshi. "The Oldest Stone Artifacts from Gona (2.6–2.5 Ma), Afar, Ethiopia: Implications for Understanding the Earliest Stages of Stone Knapping." In *The Oldowan: Case Studies into the Earliest Stone Age*, edited by Nicholas Toth and Kathy Schick, 43–75. Gosport, Ind.: Stone Age Institute Press, 2006.

Shaw, George Bernard. *Back to Methuselah: A Metabiological Pentateuch*. New York: Brentano, 1921.

———. "Maxims for Revolutionists." In *Man and Superman*. New York: Brentano, 1905.

Shea, John J. "Middle Paleolithic Spear Point Technology." In *Projectile Technology*, edited by Heidi Knecht, 79–106. New York: Plenum Press, 1997.

———. "Neandertals, Competition, and the Origin of Modern Human Behavior in the Levant." *Evolutionary Anthropology* 12 (2003): 173–187.

Shelley, Mary. *Frankenstein*. New York: Random House, 1993.

Sherwood, Chet S., James K. Rilling, Ralph L. Holloway, and Patrick R. Hof. "Evolution of the Brain in Humans—Comparative Perspective." In *Encyclopedia of Neuroscience*, edited by Mark D. Binder, Nobutaka Hirokawa, and Uwe Windhorst, 1334–1338. New York: Springer, 2009.

Shlain, Leonard. *Art and Physics: Parallel Visions in Space, Time, and Light*. New York: Morrow, 1991.

Shovkoplyas, I. G. "Dobranichevskaya Stoyanka na Kievshchine." *Materialy i Issledovaniya po Arkheologii SSSR* 185 (1972): 177–188.

Simpson, George Gaylord. *The Major Features of Evolution*. New York: Simon and Schuster, 1953.

Sinitsyn, Andrei A. "Nizhnie Kul'turnye Sloi Kostenok 14 (Markina Gora) (Raskopki 1998–2001 gg.)." In *Kostenki v Kontekste Paleolita Evrazii*, edited by Andrei A. Sinitsyn, V. Ya. Sergin, and John F. Hoffecker, 219–236. St. Petersburg: Russian Academy of Sciences, 2002.

Smail, Daniel Lord. *On Deep History and the Brain*. Berkeley: University of California Press, 2008.

Soffer, O., J. M. Adovasio, J. S. Illingsworth, H. A. Amirkhanov, N. D. Praslov, and M. Street. "Palaeolithic Perishables Made Permanent." *Antiquity* 74 (2000): 812–821.

Solecki, Ralph S. *Shanidar, the First Flower People*. New York: Knopf, 1971.

Sollas, William J. *Ancient Hunters and Their Modern Representatives*. London: Macmillan, 1911.

Sommer, Jeffrey D. "The Shanidar IV 'Flower Burial': A Re-evaluation of Neanderthal Burial Ritual." *Cambridge Archaeological Journal* 9 (1999): 127–129.

Spencer, Herbert. *The Principles of Psychology*. 3rd ed. New York: Appleton, 1881.

Stone, Valerie E. "Footloose and Fossil-Free No More: Evolutionary Psychology Needs Archaeology." *Behavioral and Brain Sciences* 25 (2002): 420–421.

Stout, Dietrich, and Thierry Chaminde. "The Evolutionary Neuroscience of Tool Making." *Neuropsychologia* 45 (2007): 1091–1100.

Stout, Dietrich, Nicholas Toth, Kathy Schick, Julie Stout, and Gary Hutchins. "Stone Tool-Making and Brain Activation: Positron Emission Tomography (PET) Studies." *Journal of Archaeological Science* 27 (2000): 1215–1223.

Straus, Lawrence G. "The Upper Paleolithic of Europe: An Overview." *Evolutionary Anthropology* 4 (1995): 4–16.

Striedter, Georg F. *Principles of Brain Evolution*. Sunderland, Mass.: Sinauer, 2005.

Stringer, Chris, and Peter Andrews. *The Complete World of Human Evolution*. New York: Thames and Hudson, 2005.

Stutz, Aaron Jonas, Natalie D. Munro, and Guy Bar-Oz. "Increasing the Resolution of the Broad Spectrum Revolution in the Southern Levantine Epipaleolithic (19–12 ka)." *Journal of Human Evolution* 56 (2009): 294–306.

Suddendorf, Thomas, Donna Rose Addis, and Michael C. Corballis. "Mental Time Travel and the Shaping of the Human Mind." *Philosophical Transactions of the Royal Society B* 364 (2009): 1317–1324.

Susman, Randall L. "Hand of *Paranthropus robustus* from Member 1, Swartkrans: Fossil Evidence for Tool Behavior." *Science* 240 (1988): 781–784.

Susman, Randall L., Jack T. Stern, and William L. Jungers. "Locomotor Adaptations in the Hadar Hominids." In *Ancestors: The Hard Evidence*, edited by Eric Delson, 184–192. New York: Liss, 1985.

Svoboda, Jiří, Vojen Ložek, and Emanuel Vlček. *Hunters Between East and West: The Paleolithic of Moravia*. New York: Plenum Press, 1996.

Swanson, Larry W. *Brain Architecture: Understanding the Basic Plan*. Oxford: Oxford University Press, 2003.

Tattersall, Ian. *The Fossil Trail: How We Know What We Think We Know About Human Evolution*. New York: Oxford University Press, 1995.

——. "Human Origins: Out of Africa." *Proceedings of the National Academy of Sciences* 106 (2009): 16018–16021.

Temple, Robert. *The Genius of China: 3,000 Years of Science, Discovery, and Invention*. New York: Simon and Schuster, 1986.

Thomas, Lewis. *The Lives of a Cell: Notes of a Biology Watcher*. New York: Viking Press, 1974.

Tobias, Philip V. *The Brain in Hominid Evolution*. New York: Columbia University Press, 1971.

Toth, Nicholas. "The Oldowan Reassessed: A Close Look at Early Stone Artifacts." *Journal of Archaeological Science* 12 (1985): 101–120.

Toth, Nicholas, Kathy Schick, E. Sue Savage-Rumbaugh, Rose A. Sevick, and Duane M. Rumbaugh. "*Pan* the Toolmaker: Investigations into the Stone Tool-Making and Tool-Using Capabilities of a Bonobo (*Pan paniscus*)." *Journal of Archaeological Science* 20 (1993): 81–91.

Toth, Nicholas, Kathy Schick, and Sileshi Semaw. "A Comparative Study of the Stone Tool-Making Skills of *Pan*, *Australopithecus*, and *Homo sapiens*." In *The Oldowan: Case Studies into the Earliest Stone Age*, edited by Nicholas Toth and Kathy Schick, 155–222. Gosport, Ind.: Stone Age Institute Press, 2006.

Toynbee, Arnold. *The Industrial Revolution*. Boston: Beacon Press, 1956.

Trigger, Bruce G. "Childe's Relevance to the 1990s." In *The Archaeology of V. Gordon Childe*, edited by David R. Harris, 9–34. Chicago: University of Chicago Press, 1994.

——. *Gordon Childe: Revolutions in Archaeology*. New York: Columbia University Press, 1980.

——. *A History of Archaeological Thought*. Cambridge: Cambridge University Press, 1989.

Trinkaus, Erik. "Neanderthal Limb Proportions and Cold Adaptation." In *Aspects of Human Evolution*, edited by Chris Stringer, 187–224. London: Taylor & Francis, 1981.

Tuchman, Barbara W. *A Distant Mirror: The Calamitous 14th Century*. New York: Knopf, 1978.

Turing, Alan. "Computing Machinery and Intelligence." *Mind* 59 (1950): 433–460.

Tuttle, Russell H. *Apes of the World: Their Social Behavior, Communication, Mentality and Ecology*. Park Ridge, N.J.: Noyes, 1986.

Valde-Nowak, Pawel, Adam Nadachowski, and Mieczyslaw Wolsan. "Upper Palaeolithic Boomerang Made of a Mammoth Tusk in South Poland." *Nature* 329 (1987): 436–438.

Vandiver, Pamela P., Olga Soffer, Bohuslav Klima, and Jiří Svoboda. "The Origins of Ceramic Technology at Dolni Věstonice, Czechoslovakia." *Science* 246 (1989): 1002–1008.

Vanhaeran, Marian, and Francesco d'Errico. "Aurignacian Ethno-linguistic Geography of Europe Revealed by Personal Ornaments." *Journal of Archaeological Science* 33 (2006): 1105–1128.

Vanhaeran, Marian, Francesco d'Errico, Chris Stringer, Sarah L. James, Jonathan A. Todd, and Henk K. Mienis. "Middle Paleolithic Shell Beads in Israel and Algeria." *Science* 312 (2006): 1785–1788.

Van Peer, Philip. *The Levallois Reduction Strategy*. Madison, Wis.: Prehistory Press, 1992.

Veblen, Thorstein. *The Instinct of Workmanship and the State of the Industrial Arts*. New York: Macmillan, 1914.

Villa, Paola. "Middle Pleistocene Prehistory in Southwestern Europe: The State of Our Knowledge and Ignorance." *Journal of Anthropological Research* 47 (1991): 193–217.

Villa, Paola, François Bon, and Jean-Christophe Castel. "Fuel, Fire and Fireplaces in the Palaeolithic of Western Europe." *Review of Archaeology* 23 (2002): 33–42.

Villa, Paola, S. Soriano, Nicolas Teyssandier, and S. Wurz. "The Howiesons Poort and MSA III at Klasies River Main Site, Cave 1A." *Journal of Archaeological Science* 37 (2010): 630–655.

Wadley, Lyn. "Revisiting Cultural Modernity and the Role of Red Ochre in the Middle Stone Age." In *The Prehistory of Africa: Tracing the Lineage of Modern Man*, edited by Himla Soodyall, 49–63. Johannesburg: Jonathan Ball, 2006.

——. "What Is Cultural Modernity? A General View and a South African Perspective from Rose Cottage Cave." *Cambridge Archaeological Journal* 1 (2001): 201–221.

Walker, Alan, and Richard E. Leakey. "The Postcranial Bones." In *The Nariokotome Homo erectus Skeleton*, edited by Alan Walker and Richard E. Leakey, 95–160. Berlin: Springer, 1993.

Walker, Alan, and Pat Shipman. *The Ape in the Tree: An Intellectual and Natural History of Proconsul*. Cambridge, Mass.: Belknap Press of Harvard University Press, 2005.

——. *The Wisdom of the Bones: In Search of Human Origins*. New York: Knopf, 1996.

Watts, Ian. "The Origin of Symbolic Culture." In *The Evolution of Culture: An Interdisciplinary View*, edited by Robin Dunbar, Chris Knight, and Camilla Power, 113–146. New Brunswick, N.J.: Rutgers University Press, 1999.

Weaver, Timothy D. "The Meaning of Neandertal Skeletal Morphology." *Proceedings of the National Academy of Sciences* 106 (2009): 16028–16033.

Wells, Spencer. *Deep Ancestry: Inside the Genographic Project*. Washington, D.C.: National Geographic Society, 2007.

Westermarck, T., and E. Antila. "Diet in Relation to the Nervous System." In *Human Nutrition and Dietetics*, 10th ed., edited by J. S. Garrow, W. P. T. James, and A. Ralph, 715–730. New York: Churchill Livingstone, 2000.

Wheeler, William Morton. "The Ant-Colony as an Organism." *Journal of Morphology* 22 (1911): 307–325.

White, Lynn. *Medieval Religion and Technology: Collected Essays*. Berkeley: University of California Press, 1978.

——. *Medieval Technology and Social Change*. London: Oxford University Press, 1962.

White, Mark, and Nick Ashton. "Lower Palaeolithic Core Technology and the Origins of the Levallois Method in North-Western Europe." *Current Anthropology* 44 (2003): 598–609.

White, Tim D., Berhane Asfaw, Yonas Beyene, Yohannes Haile-Selassie, C. Owen Love-joy, Gen Suwa, and Giday WoldeGabriel. "*Ardipithecus ramidus* and the Paleobiology of Early Hominids." *Science* 326 (2009): 75–86.

White, Tim D., Gen Suwa, and Berhane Asfaw. "*Australopithecus ramidus*, a New Species of Early Hominid from Aramis, Ethiopia." *Nature* 371 (1994): 306–312.

Willoughby, Pamela R. *The Evolution of Modern Humans in Africa: A Comprehensive Guide.* Lanham, Md.: AltaMira Press, 2007.

Wilson, Andrew I. "Machines in Greek and Roman Technology." In *The Oxford Handbook of Engineering and Technology in the Classical World*, edited by John Peter Oleson, 336–366. Oxford: Oxford University Press, 2008.

Wilson, Edward O. *The Insect Societies.* Cambridge, Mass.: Belknap Press of Harvard University Press, 1971.

Wilson, Frank R. *The Hand: How Its Use Shapes the Brain, Language, and Human Culture.* New York: Pantheon Books, 1998.

Wood, Bernard. *Human Evolution: A Very Short Introduction.* Oxford: Oxford University Press, 2005.

Wulliman, Mario F. "Brain Phenotypes and Early Regulatory Genes: The *Bauplan* of the Metazoan Central Nervous System." In *Brain Evolution and Cognition*, edited by Gerhard Roth and Mario F. Wulliman, 11–40. New York: Wiley, 2001.

Wymer, John. *Lower Palaeolithic Archaeology in Britain as Represented by the Thames Valley.* London: Baker, 1968.

Wynn, Thomas. "Archaeology and Cognitive Evolution." *Behavioral and Brain Sciences* 25 (2002): 389–438.

——. "The Evolution of Tools and Symbolic Behavior." In *Handbook of Human Symbolic Evolution*, edited by Andrew Lock and Charles R. Peters, 269–271. Oxford: Blackwell, 1999.

——. "Handaxe Enigmas." *World Archaeology* 27 (1995): 10–24.

——. "Piaget, Stone Tools and the Evolution of Human Intelligence." *World Archaeology* 17 (1985): 32–43.

——. "Tools, Grammar and the Archaeology of Cognition." *Cambridge Archaeological Journal* 1 (1991): 191–206.

Wynn, Thomas, and Frederick L. Coolidge. "Implications of a Strict Standard for Recognizing Modern Cognition in Prehistory." In *Cognitive Archaeology and Human Evolution*, edited by Sophie A. de Beaune, Frederick L. Coolidge, and Thomas Wynn, 117–127. Cambridge: Cambridge University Press, 2009.

Wynn, Thomas, and Forrest Tierson. "Regional Comparison of the Shapes of Later Acheulean Handaxes." *American Anthropologist* 92 (1990): 73–84.

Zebrowski, Ernest. *A History of the Circle: Mathematical Reasoning and the Physical Universe.* New Brunswick, N.J.: Rutgers University Press, 1999.

Zhu, R. X., R. Potts, Y. X. Pan, H. T. Yao, L. Q. Lü, X. Zhao, X. Gao, L. W. Chen, F. Gao, and C. L. Deng. "Early Evidence of the Genus *Homo* in East Asia." *Journal of Human Evolution* 55 (2008): 1075–1085.

INDEX

Darwin, Charles, 8, 13–14, 19, 23, 42, 106, 172, 178, 182n.41. *See also* natural selection

Darwin, Erasmus, 175

Davidson, Iain, 192n.100, 200n.64, 202n.85. *See also* archaeology: cognitive

Dawkins, Richard, 15

Dawn of European Civilization, The (Childe), 104. *See also* archaeology: culture concept of

Deacon, Terrence W., 198n.42. *See also* language, syntactic

death, denial of, 121. *See also* intentionality

Decline of the West, The (Spengler), 106. *See also* history

Deetz, James, 31, 97, 186n.1, 199n.48. *See also* language, syntactic: artifacts and; mental template

dendrites, 36. *See also* brain; neurons

Descartes, René, 13, 165, 172. *See also* mind–body dualism

Dessauer, Frederick, 22. *See also* technology: philosophy of

digital, 36, 87. *See also* analogical; language, syntactic; representations, mental; symbols

digging implements, 101, 118, 149

discrete infinity, 6, 74, 182n.43. *See also* Chomsky, Noam; creativity; language, syntactic

Dmanisi (Georgia), 62. *See also* Oldowan industry

DNA: ancient, 32, 195n.14; mitochondrial, 132; of Neanderthals, 195n.14

Dobranichevka (Ukraine), 133, 135. *See also* dwelling: of mammoth bones

dog. *See* domestication: of Upper Paleolithic dog

Dolní Věstonice (Czech Republic), 123, 127. *See also* Gravettian industry

domestication, 129, 132, 145; of donkey, 149–150; of horse, 156–157; of ox, 149; of plants, 145; of Upper Paleo-

lithic dog, 129, 132, 210n.96. *See also* agriculture

Donald, Merlin, 16, 27, 185n.79. *See also* symbols: storage of

Dorset, Late (arctic culture), 168

drill, rotary, 101, 118. *See also* Kostenki-Borshchevo sites; Upper Paleolithic: early

Dryopithecus, 41, 45, 50, 188n.37. *See also* apes; hand, human: evolution of

dualism. *See* mind–body dualism

Dunbar, Robin, 20, 83. *See also* language, syntactic; "social brain hypothesis"

dwelling (house), 120, 133, 135–137, 145; of mammoth bones (late Upper Paleolithic), 133, 135; paved floor in (late Upper Paleolithic), 135–136; rectilinear, 145. *See also* 'Ain Ghazal; Dobranichevka; Kostenki-Borshchevo sites; Mezhirich; Yudinovo

E. coli (bacterium), 35. *See also* brain: evolution of

Edelman, Gerald, 21. *See also* brain; consciousness; language, syntactic; mind, human

Egypt, ancient, 151–152, 154–156; Akhenaten, 154; calendar in, 152; Cheops, 151; Djoser's step pyramid in, 155; Fourth Dynasty of, 151; Great Pyramid in, 151, 153; longevity of, 155; Middle Kingdom "police state" in, 156, 215n.53; nation-state imposed on clans and villages in, 155–156; *nomes* in, 156, 215n.52; Old Kingdom of, 151, 154, 156, 158; *sed* festival in, 155, 215n.50; Sixth Dynasty of, 156; technology in, 215n.49; Third Dynasty of, 155; tombs as art in, 212n.11; Twelfth Dynasty of, 156; unification of Lower and Upper, 152; writing in, 155, 215n.51. *See also* Assmann, Jan; China; creativity: suppression of; Kemp, Barry J.; Sumer

Eliseevichi I (Russia), 132. *See also* domestication: of Upper Paleolithic dog

Hadar (Ethiopia), 47, 50, 52, 191n.83. See also *Australopithecus afarensis*

hand, human, 2–3, 35, 43, 190n.56; apical tufts of, 50–51; of *Australopithecus afarensis*, 50–51; evolution of, 47–52, 180n.7; glabrous skin of, 48; grips of, 49–52; of *Homo ergaster*, 52, 190n.76; of *Homo habilis*, 51–52, 190n.73; movements of, 48–50; opposition of finger and thumb, 48, 50; of *Paranthropus robustus*, 51–52; ratio of thumb length to index finger length, 48, 50; as sensory organ ("sensory hand"), 48, 107, 165. See also bipedalism; Mountcastle, Vernon; Napier, John Russell; representations, mental: artificial; vocal tract, human

hand ax. *See* biface

"hard-hammer" flaking technique, 89

harpoon, 101

Hegel, G. W. F., 172–173, 178, 220n.14. *See also* history

Heidegger, Martin, 22, 25, 27. *See also* technology: philosophy of

Heinrich Event 4, 118

Hellenistic period. *See* Greece, ancient: Hellenistic period of

hemoglobin, 37

Herder, Johann, 172

Herto (Ethiopia), 81. *See also* humans, modern: anatomy of

history, 106–107, 171–174; mind and, 106, 172–174; "people without," 166–168, 219n.95; prehistory and, 104–109; as progress, 172–174; theocratic, 154; Upper Paleolithic as, 108–137. *See also* Childe, V. Gordon; Collingwood, R. G.; creativity; Hegel, G. W. F.; Kant, Immanuel; mind, human; Smail, Daniel Lord

Hitler, Adolf, 104

Hobbes, Thomas, 13, 17–20, 130, 172, 176, 182n.37; on mechanical technology as "artificiall life," 17, 176. *See also* eusociality; *Leviathan*; super-brain, modern human; super-organism

Hohle Fels Cave (Germany), 10, 120–121, 181n.29. *See also* music; visual art

Hohlenstein-Stadel (Germany), 101–102, 120. See also *Löwenmensch* figurine

Holloway, Ralph, 31. *See also* brain: evolution of

Homo erectus, 52, 61, 100

Homo ergaster, 52–53, 61–62, 65

Homo habilis, 57, 61–62, 190n.73

Homo heidelbergensis, 53, 67, 70–71, 79–81, 91

Homo neanderthalensis. *See* Neanderthals

Homo sapiens, 4, 70, 73, 79–80, 83. *See also* humans, modern

honeybee (*Apis* sp.), x, 16–17, 70, 75, 211n.105; externalized representations ("dance language") of, 16, 19, 68, 87, 179n.5. *See also* eusociality; representations, mental; super-brain, modern human; super-organism

humans, modern: anatomy of, x, 8, 29, 93; brain volume of, 80–81, 84; colonization of Australia by, 100–101, 112–113; dispersal of, out of Africa, 16, 112–115, 167, 174, 183n.50, 206n.38; evolution of, 80–83; identification of, from isolated teeth, 207n.50; as "information animal," 84; information overload and, 197n.33; self-redesign of, as organisms, 100, 113, 139, 176, 213n.18; tropical-climate adaptation of, 115, 128. See also *Homo sapiens*; language, syntactic; Middle Stone Age, in Africa; mind, human; modernity; Neanderthals; super-brain, modern human; Upper Paleolithic

hunter-gatherers, recent, 166–168; historical change among, 167

Huxley, Thomas H., 42. *See also* Darwin, Charles

hypothalamus, 37

"ice cellar." *See* storage, of perishables: cold

Idea of History, The (Collingwood), 106

Klein, Richard G.; language, syntactic; modernity

neurons, 15–16, 35–37, 65. *See also* synapses

neurotechnology. *See* technology: information

Noble, William, 192n.100, 200n.64, 202n.85. *See also* archaeology: cognitive

Notarchirico (Italy), 194n.122

notochord, 37

noumena (things-in-themselves), 22, 29. *See also* Dessauer, Frederick; Kant, Immanuel; *phenomena*; technology: philosophy of

Ohalo II (Israel), 143–144, 213n.18. *See also* agriculture; baking oven

Oldowan industry (Lower Paleolithic), 53–62, 68, 191n.78; association of, with *Homo*, 191n.83; categories of artifacts of, 55; "Developed Oldowan B" assemblages of, 57; microscopic wear on artifacts of, 56; problem of defining artifacts of, 55; "spheroids" of, 192n.99. *See also* biface; Leakey, Mary; mental template; Olduvai Gorge; proto-biface; Toth, Nicholas

Olduvai Gorge (Tanzania), 51–52, 54–60, 62, 190n.73, 191n.83; major artifact-bearing units of, 57–58. *See also* Acheulean industry; Oldowan industry

Old World Monkeys, 15, 39, 41. *See also* primates

Oligocene epoch, 39

Olorgesailie (Kenya), 68–70. *See also* Isaac, Glynn Ll.

Omo Kibish (Ethiopia), 81. *See also* humans, modern: anatomy of

On the Origin of Species (Darwin), 13, 78

optic tectum, 37–38. *See also* vision

orangutan, 42. *See also* apes

ornaments, 98–99, 113–114; of Neanderthals, 203n.90; of perforated shell,

98–99, 113–114, 203n.88. *See also* Blombos Cave; Oued Djebbana; Skhul

Orrorin turgenensis, 47. *See also* bipedalism

Oswalt, Wendell H., 219n.96. *See also* techno-units

Oued Djebbana (Algeria), 98–99

Paleozoic era, 38

Paranthropus robustus, 51–52, 54, 57. *See also* hand, human

Pasteurella pestis. See bubonic plague

Peirce, Charles Sanders, 181n.31. *See also* representations, mental; symbols

Peloponnesian War, 157. *See also* Greece, ancient

pelvis, human, 45. *See also* bipedalism

Peninj (Tanzania), 200n.55

periglacial steppe, 122–123

Persian Empire, 156

Petralona (Greece), 71, 81. See also *Homo heidelbergensis*

phenomena (things-as-they-appear), 22. *See also* Dessauer, Frederick; Kant, Immanuel; *noumena*; technology: philosophy of

Philip of Macedon, 157. *See also* Greece, ancient

Philosophy of History (Hegel), 173

phrase structure grammar, 7, 11. *See also* language, syntactic

Pinker, Steven, 14, 75, 195n.7. *See also* mind, human

Plateau Parrain (France), 137. *See also* dwelling

Platyhelminthes, 36. *See also* brain: evolution of

Pleistocene epoch (Ice Age), 128

plow, 149, 158, 161; in China, 158; iron, 158; moldboard, 161. *See also* agriculture; technology: innovation in

Pool, Robert, 25. *See also* super-brain, modern human

positron-emission tomography (PET), 27, 32, 85, 193n.102. *See also* brain-imaging technology